LIFE ON THE LINE

LIFE ON THE LINE

Young Doctors Come of Age in a Pandemic

Emma Goldberg

HARPER
An Imprint of HarperCollins*Publishers*

 For information, address HarperCollins Publishers, 195 Broadway, New York, NY 10007.

HarperCollins books may be purchased for educational, business, or sales promotional use. For information, please email the Special Markets Department at SPsales@harpercollins.com.

FIRST EDITION

Library of Congress Cataloging-in-Publication Data has been applied for.

ISBN 978-0-06-307338-8

21 22 23 24 25 LSC 10 9 8 7 6 5 4 3 2 1

For my mom, Shifra

A Note from the Author

This book is a work of narrative nonfiction, in that all events described are real and based on extensive interviews with six doctors: Sam, Gabriela, Iris, Elana, Jay, and Ben. But this is not a work of live reporting, in that the author did not observe these events firsthand; the dialogue was reconstructed from the memories shared. Jay is referred to by her nickname and her family members' names have been changed to protect their privacy. The names of her medical school and hospital have not been identified to protect the privacy of patients described in detail.

The names of all patients have been changed to protect their privacy, though the hospital convention of referring to them by their last names has been maintained. Patient ages and dates of hospital admission have not been identified. Any resemblance to persons living or dead resulting from these changes is coincidental and unintentional.

Introduction

"Next up—Samuel."

The applause sounded like it was coming from some parallel universe. One where he would have crossed a graduation stage and seen the faces of his mom and his boyfriend hollering from somewhere in the sea of seats in Alice Tully Hall in Lincoln Center. One where he could have waved his graduation cap in the air, grinned, and shuffled forward in the regal purple robe of New York University.

But instead here he was, hearing his name called and his friends cheer at 7:30 on a rainy April morning in Bellevue Hospital as he and his classmates received their staff badges. It was a full three months before any of them had thought they'd be starting work. They stood a couple of feet from one another, whoops muffled by masks. Their group seemed impossibly small in that cavernous space, normally bustling with throngs of patients, nurses, visitors, and homeless New Yorkers.

Sam came forward and took the ID out of the outstretched hand of a hospital administrator. There was his name, printed alongside the block letters MD.

"Woo-hoo!" His classmates clapped as Sam rejoined their ranks.

Deprived by their weeks of quarantine of any real graduation ceremony except for a short event on video, some of Sam's classmates were approaching their first day on the job, in Bellevue's coronavirus wards, as a pseudo-celebration. This was the day they'd looked forward to for years—the day they would finally feel useful, no longer underfoot.

Every minute that morning held a sense of anticipation, of the strange stillness preceding a storm. There was Sam's commute from Greenwich Village on an empty M23 bus. There was the ghostly quiet of the block around Bellevue. Then there was the line of doctors in scrubs getting their temperatures taken with a temporal artery thermometer as they entered the building, the familiar faces tougher to pick out behind their masks. Someone in a face shield took Sam's temperature and pronounced it normal.

With their badges in hand, the new doctors lined up to take Bellevue's notoriously slow elevator upstairs, where they'd be fitted for protective equipment. They traded tips and grievances about the endless preparations for their actual residencies, set to start in July after this temporary Covid-19 assignment. After three hours of paperwork, their orientation was complete and they were free to head home. It was a straightforward affair. The next morning at 7:00 they would report for their first handoff, which meant reviewing the list of patients that they were responsible for.

As Sam headed toward the exit, the rain outside was falling in sheets. The city streets sat limp, like laundry waiting to be wrung out.

It was April 13. New York State would lose 778 more patients to coronavirus before the start of Sam's first hospital shift the next morning.

There are few relationships more intimate than the one between a patient and doctor. The stakes are not just the heart but the living self, and the dependence can be absolute. It is an act of total trust to submit one's body to another's hands.

It's often said that doctors play God, that they determine life or death through the medical orders they issue. But of course they aren't all-powerful or particularly mysterious. They are humans, pressing through a long shift, rolling their ankles to relieve the tension of a day on their feet, mulling over the takeout they'll order for dinner, hoping their supervisor caught their deft insertion of an IV needle. They are normal people with an outsize influence on how we live or die.

Yet their impact on our lives is powerful. It is no wonder that they are granted an unusual degree of societal status—while the median income for an American adult is roughly $50,000 a year, for a physician it is four times that amount. But medicine distinguishes itself from other high-paying fields by its alchemy of conscience, merit, and money. It is one of the last remaining affluent professions that stakes a claim to earning its pay fairly: in raw talent, unflagging exertion, and a clear-eyed commitment to public good.

In its devotion to elitism masked as meritocracy, the story of America's doctors mirrors the story of this country. The medical profession has rendered itself exclusive by design. In the late nineteenth and early twentieth centuries, medical organizations surveyed the field and decided it wasn't quite restrictive enough, so they undertook an effort to close medical schools, shrink the number of annual graduates, and ensure that those that remained met a standard of rigorous performance. The effect was also to make those school populations more affluent, more male, and overall more homogeneous. A century later, American medical schools are still predominantly white and wealthy. This is partly because doctors are predominantly white and wealthy, and doctors tend to beget doctors.

In recent decades, many medical schools have made a concerted effort to draw their student bodies from a wider range of racial and socioeconomic backgrounds. Some have come to realize that such diversity is integral to the work. Quite simply, it improves the health of patients. Recent studies—the type that might not have been funded just a few decades ago—show that Black patients have better outcomes when treated by Black doctors; they are more likely to bond with these doctors emotionally and agree to important preventive care measures like cholesterol tests and diabetes screenings. It's likely, too, that they're less apt to face racial bias.

In other words: in the exam room, identity matters, both the doctor's and the patient's.

And so the field's exclusive bounds have begun to crack. As the medical profession has started to look more like its patients, a shift has emerged, not just in the culture surrounding the work but also in the outcomes of that work. When patients and doctors see themselves in one another, they forge stronger bonds and have more candid conversations. They talk more openly about preventive care screenings, medications, and which invasive measures to forgo at the end of life. They build the type of trust that has long been absent from clinical examination rooms.

Gradually, the face of American medicine has been changing. It has been growing less homogeneous. It has also been becoming more open, less distant from patients, and more focused on communication and connection. And then, on top of these two slow shifts, there came a once-in-a-lifetime crisis.

In March 2020, a virus far smaller than a grain of sand upended America's medical institutions. With hospitals under siege, a handful of medical schools fast-tracked graduation and sent their fourth-year students to the front lines. These newly minted doctors immediately saw their field's inequities laid bare. The people most likely to get sick and die from Covid-19 were the country's most vulnerable: nonwhite and working-class people. African American and Hispanic people were hospitalized at four times the rate of white Americans. They were also more likely to have essential jobs that didn't allow them to socially isolate. The country's deep-rooted inequalities, made manifest in its health disparities, became crystalline to the new doctors as they saw Black bodies filling their hospital beds.

At the same time, the recruits also saw work norms shaken—sometimes for the better. Hospital teams found new mechanisms for collaboration. Typically, medicine holds its hierarchies sacred, but in a time of crisis what matters most is the shared effort that helps the work get done. The fear of infection forced doctors to minimize time at bedsides, so they had to rethink their communication with patients. Some doctors also felt closer than ever to their patients, who saw them risking everything to offer care.

The doctors at the center of this narrative represent the new face of American medicine. Gabriela, a young Hispanic woman, was raised by a mother who did not go to college. Iris is a first-generation American, whose Chinese parents practice traditional medicine. Sam, who is gay, resolves to join the front lines of a modern-day plague, knowing the stories of people failed by the health care system during the AIDS epidemic. Jay eschews the models of cold authority she saw in older physicians in favor of a more human approach to caring for her patients. Aware of the health care profession's history of going to any lengths to stave off death, Ben was drawn to medicine by an interest in helping patients determine how they want to die. And Elana would learn viscerally what it means to treat patients like family when a member of her own family fell sick.

All in their twenties, these young doctors were raised with a traditional image of medicine—as a profession of power, stature, exacting standards. They understood that their chosen profession came laden with cultural and social weight. But their interest in it was shaped by something humbler: a desire to serve people who look and grew up like them. Immigrants who fear American institutions like hospitals. People of color who were taught to distrust the medical system. Gay people who had their identities pathologized. These doctors know those experiences. They know the fear that white coats and sterile white halls can so easily provoke. They are intent on remaking the field in their image.

For the newest doctors, the frontline Covid-19 months have brought a dual sense of obligation and uncertainty. Their commitments to forging trust and caring for the marginalized have felt more urgent than ever. Those same commitments have also felt impossibly hard.

For all the things we spend our days talking about—politics, art, fashions, love, and faith—there's nothing more fundamental than our bodies. Covid-19 has served as a visceral reminder of that. In putting our health at the forefront of our minds and our doctors at the center of national conversation, the pandemic could have far-reaching effects on the medical profession. Some, we won't understand for years to come, but the experiences of those who served on the front lines hold some clues. This is the story of unusual doctors who entered the field at an unusual time.

I once would have loved to be a doctor, but after spending much of high school chemistry squinting at the board in confusion, I decided the humanities were more my speed. Shortly after I started working at the *New York Times*, however, I pitched a story on the barriers to entering medicine, particularly for low-income students. I ended up interviewing dozens of students who came from demographics that never saw themselves represented in the health care system: Hispanic students, undocumented students, students whose parents didn't go to college. What I saw was a profession in the midst of a tectonic shift, opening up and breaking its boundaries down. I saw a field beginning to confront its long history of bias and exclusion. I became increasingly interested in the culture of medicine, all the relationships and social forces that affect the way patients receive care.

When medical schools across New York and Massachusetts decided to graduate their students early to help hospitals besieged by the coronavirus, I reported the story for the *Times*. I then began following some of those New York students, speaking with them daily about their experiences. As my Brooklyn neighborhood filled with sirens and my Twitter feed collapsed into gloom, I found solace in these daily phone calls. Here were people, my own age, setting aside their fears to ensure that our city and their profession could rebuild. I am fortunate that these doctors gave me their time day after day, even as they worked ten-hour shifts in the hospitals. They called on lunch breaks and on subway platforms and late at night from bed. They shared their notes and photos. They even called when I was worried I'd gotten sick. They connected me to their family members, partners, and friends. Over months of interviews, I pieced together what they saw.

Sometimes we mused together, during these calls, over how we'd make sense of this time. We questioned how writers and artists and future historians would capture what it felt like to see our lives mutate overnight because of something totalizing and invisible. We hoped they would show how the changes were felt most in the smallest moments, grocery runs and calls home to Mom. We wondered: What would the historical record look like? Throughout New York's long, awful Covid spring, I spoke with ordinary people doing extraordinary work; I hope their stories can serve as one chapter.

One

SAM, March 2020

The third Friday in March was Match Day. After four years of medical school, Sam and his classmates would be placed into residency programs for the next phase of their training. For Sam, the day held extra significance. His parents were supposed to meet his boyfriend Jeremy's family—one of those gatherings that bore the peculiar quality of familiarity and awkward novelty all at once. Sam's family would fly in from Ohio, Jeremy's from Texas.

Match Day is when you find out which of the hospitals you've listed as your top choices has picked you too. It is the archetypal medical coming-of-age rite. On Match Day, the next three years of Sam's life would take on distinct shape, contoured by the program and city where he would train (which he hoped would be at NYU).

The NYU match ceremony, which just Sam's family would attend, was scheduled for Friday afternoon. Sam and Jeremy agreed that their families' first meeting would be later that night, over Sabbath dinner at their apartment. They'd both grown up in Jewish families that marked Friday evenings with too much food and boisterous conversation. They knew there was no weighty interaction that the right menu couldn't ease, and settled accordingly on handmade pasta from Raffetto's, the luxe Italian grocery store two blocks from their apartment, paired with a classic Trader Joe's spring mix salad.

If you zoomed out as though to see them through the long end of a telescope, just the faint edges of their lives visible, you'd glimpse a little Greenwich Village apartment as distinctly New York as a Noah Baumbach film. Two boys from Cincinnati and Fort Worth, bodies entangled on the couch. Jeremy is slim, with a square jawline and closely cropped brown hair. His face is small and his features delicate, which makes his big, warm eyes seem to take up even more space. He is well dressed, in a varsity jacket, a heather-gray T-shirt, and skinny jeans. Sam is slightly taller and broader-shouldered, his mouth fixed in a permanently impish grin. He has a crew cut and thick, dark brows, a silver piercing in his ear, and an ever-present layer of scruff on his chin.

Sam and Jeremy met marching in the New York City AIDS walk for their synagogue, a queer congregation called Beit Simchat Torah in North Chelsea. When they moved in together in Sam's third year of med school, their life took on the firm architecture of a real relationship. Not the tentative will-they-won't-they of mid-twenties dating in New York, the stomach-squeezing angst of unanswered texts, but the concrete quality of partnership. They shared a home. They divvied up chores and grocery trips. And on most days they balanced one another out; Sam could be more cerebral and wry, while Jeremy was sweet and earnestly enthusiastic. He had a tone of voice that made you think anything you'd just said was a good idea.

Match Day was a validation of all that early adulthood work. Sam would find out where he would spend the next three years, his first phase as an MD. His family and Jeremy's family would meet. Names would be put to faces. Conversation would bump along, then lift off. Inside jokes would take form, becoming part of their families' shared history.

Like so much of the news around coronavirus, the sense of loss came gradually and then all at once. Sam got a call from his parents one afternoon. "What do you think will happen with Match Day?" they asked him. "How is this all going to work?" Sam's mom was a planner. She was probably scanning the terms of the flights that she'd booked months ago on the other end of the line.

"I'm not sure," Sam said. They were all holding out hope that it would happen, though this was partly shaped by the desire to cling to all the events surrounding Match Day, the Shabbat dinner and the family vacation to the Grand Canyon they had planned for the spring.

Sometime later, Sam got an email from NYU: the ceremony would be limited just to graduating class, faculty, and staff. Sam's parents soon realized there would be no flight from Cincinnati to New York.

In the days that followed, the other normalities of their life evaporated so quickly that by the time the announcement came that Match Day would be virtual, it was no surprise. The NBA shut down; Broadway theaters went dark. By then, Sam and Jeremy's once-full routine had condensed to near comic emptiness. Where their days had once been a jumble of alarm clocks and MetroCards, now it was just the two of them ordering takeout to the apartment. Every trip outside, even just for milk and eggs, felt fraught. The days took on a colorless congruity; Monday was the same as Tuesday, and Tuesday the same as Saturday. The apartment also became Jeremy's office when his media company sent everyone to work from home.

On Match Day, Jeremy ran to the corner liquor store to buy champagne. Then he ran over to Posh Pop Bakery, the little shop on Bleecker that had miraculously stayed open as the rest of the city shut down, and stubbornly purchased just as many pastries as he would have if both his and Sam's families were in town. Sam didn't bother turning on the virtual ceremony; he was too nervous about actually getting his match. He FaceTimed his parents and maniacally hit refresh on his email.

It came at 11:58 a.m. "It's NYU!" It was the hospital's primary-care internal medicine program, Sam's top choice. Jeremy popped the bottle. And for a moment there was just the soundtrack of an otherworldly celebration—the cheers of his mom and dad on the phone, the hiss of champagne, and somewhere far away the blare of the city sirens.

After all the stress of medical school—the late-night cramming, the ungodly early wake-ups, the Step 1 exam, the Step 2 exam, all those endless exams—this was supposed to be a spring of unwinding, before residency started in July. Sam was going to brush up on his medical Spanish, but he and Jeremy could also do what other less busy couples did. What did other twenty-something couples do? Bake sourdough? They could bake sourdough. Some of Sam's classmates called this term, their second half of fourth year, "the most expensive vacation you'll ever pay for." But this vision of a relaxing few months was quickly evaporating.

The coronavirus had bulldozed its way through cities and populations and headlines and cable news, and by now some of its early myths were fading. It was clear that this was not just a disease of the old and sick. New York wouldn't contain it, couldn't contain it. And many people would die.

NYU decided to make an offer: Sam and his classmates could graduate a month early, if they wanted to, and work for several weeks in NYU-affiliated hospitals overwhelmed by the surge of Covid patients. They would be sent into the hospitals where they had done their advanced med school rotations, meaning Sam would go to Bellevue. Regardless of these early graduates' specialties, whether pediatrics or psychiatry, they would help in the internal medicine units, which were most feeling the crunch.

Confined to their apartment and reading apocalyptic news stories from Milan and Wuhan, Sam and Jeremy felt the weight of the situation. To Sam, starting his hospital work three months early to help tend to Covid patients seemed like a responsible choice. He was young and healthy. This was the point of his medical education: to be of use in times of crisis. He'd even sent emails to his research mentor asking about opportunities to volunteer. But both he and Jeremy realized the risk. There was so little known about the virus. Some of his classmates whispered worriedly about the infection rates among health care workers in China, where more than three thousand medical workers had been sickened with Covid.

Besides all that, Sam and Jeremy were young queer men raised just after the HIV crisis. Much like the children of Holocaust survivors who were born years after the war but inherited trauma from the camps, people who came out as gay in the 2000s inherited their own kind of anguish. Theirs were the films and the songs and the diatribes of the AIDS epidemic. All those photos of people who died in their early twenties, groups of friends wiped out in San Francisco and Chicago and New York. Right where Sam and Jeremy now lived, in the Village, whole blocks of boys had vanished. The most painful thing was thinking about the way that virus lodged itself in relationships, like a cudgel. If you were sick, then your partner was likely sick; if you took a risk, then you could kill the person you loved.

Sam and Jeremy came out in an age of condoms and non-abstinence sex education. In his sophomore year, Sam played in the ensemble band for a performance of *Rent* at his high school, which was one of the first in the country to stage the show. During college, he worked at a campus wellness center testing students for chlamydia, gonorrhea, and HIV. He even gave a TEDx talk to classmates on destigmatizing preventive health measures like taking PrEP to prevent HIV infection.

By virtue of the decade he was born in, Sam wouldn't have to carry the pain of losing his whole community, or the all-consuming fear of losing his own life. Sam's would be the generation that lived without a death sentence.

Until coronavirus. All those years of fervently preaching safe sex, and here Sam was in danger from an unknown disease. And worse still, possibly putting Jeremy in harm's way. Sam knew that he could, more likely than not, be exposed to coronavirus. Its viral particles would be everywhere as Sam leaned over patient beds to draw blood or take medical histories. The contaminated elastic of his mask could brush against his eyes as he pulled it off.

One evening they talked out their options. Jeremy's parents had just called in a panic to ask if he had considered moving out of the apartment while Sam was working at Bellevue. The two discussed it as Sam cooked dinner and Jeremy lay on the couch nearby.

"I could apartment-sit," Jeremy mused. He had a coworker with a studio a few blocks away who was staying with a boyfriend in Brooklyn.

"There are those hotels offering free rooms for health care workers," Sam said. "I think the Four Seasons in Midtown is doing it."

But Sam didn't want to spend the next two months alone; neither did Jeremy. They wanted to debrief their days and watch *Tiger King* and cook together.

They agreed that Sam would develop an elaborate routine to douse himself in sanitizer and try to prevent any possibility of taking the virus home. They said they would see how it went for the first few days and then reevaluate. But Sam knew, realistically, that he wouldn't weather the weeks ahead on his own. The most insidious part of the virus was not just the dread it sparked, but that its nature meant you might have to face that dread alone.

"Born This Way" by Lady Gaga hit number one on the Billboard Hot 100 when Sam was in high school. Sam had been spending his summers surrounded by the hormone-frenzied stealth make-outs of Jewish camp. On Friday nights the girls waited for the boys they liked to ask them on strolls around the campground. There was even a point system for hookups, with each camper vying to end the session with the highest score. But suddenly it was 2011 and Sam's classmates were dancing to a queer pride anthem: "It doesn't matter if you love him, or capital H-I-M."

The problem was that Sam didn't like the synthetic sugar pop lyrics, this marketing effort to make his queer identity into something biological and self-pathologizing. "Born This Way" was ostensibly about gay pride—but it offered only a narrow ledge to stand on. It was one step away from implying that queer identity was a clinical aberration. Which could then imply that coming out was just an acknowledgment of something genetic, a diagnosis. Where was the empowerment in that?

By then, Sam had the feeling he might want to work in medicine or biomedical research. His dad, Neil, worked as a psychiatrist and his mom, Laurie, as an occupational therapist. They had noticed his aptitude for science early on. When he was three or four, Sam used to stand in the driveway, bent over, studying the ground, and when they asked him what he was doing he said he was examining the ants. In eighth grade Sam won one of ten highly coveted spots for Cincinnati public school students to attend a sea camp in Florida funded by Jimmy Buffett. Then there was the time Sam was named Hamilton County Recycler of the Year. So Sam's parents weren't surprised when he started entering state science fairs, winning awards.

Sam began telling people he was gay his second weekend of college, at Ohio State. The next year he declared a minor in sexuality studies, and later started working at a clinic for queer health. When he applied to medical school—the year after his college graduation, while he was working as a busboy at a gourmet taco joint—he wrote on his applications that he was interested in sexual and LGBTQ health. But in interviews he found himself in the awkward position of having to out himself to the solemn-faced, white-coated figures determining his admission.

"So tell me more about why you're interested in LGBTQ health."

And Sam would sit there wondering why it was that he should have to describe his sexual orientation to the powerful people tasked with shaping his professional future. On the one hand, he wondered if directly coming out in the interview subjected him to unnecessary judgment; on the other hand, if he didn't, then was he being inauthentic? At some schools the questions were phrased in even more unsettling ways: "So what exactly is LGBTQ health?" Which could sound like an implicit suggestion that this wasn't a serious or "real" area of medicine.

For Sam it had always been about care for people on the margins. People who grew up with the nagging sense that something about them wasn't normal, that they were second-class. There were different strands of Sam's identity informing that sense of empathy. His parents were both raised in Deep South towns—Laurie in Jackson, Mississippi, and Neil in Kingstree, South Carolina—at a time when Jews were made to know in no uncertain terms that they didn't belong. Laurie's temple was bombed by the Klan when she was ten. Neil's dad used to tell him where he fit in the neighborhood hierarchy: there were the WASPs at the top, then the Catholics and Jews, and then their Black neighbors.

For Sam, becoming a doctor meant earning the authority to tell people they weren't medically wrong, that science shouldn't be used as a weapon against them. He could tell them all the ways that they'd been let down by systems and by rigid social norms, and how they would survive regardless.

Neil and Laurie were proud that all the values they'd instilled in Sam—the importance of Being Good and Doing Unto Your Neighbor—were now the fibers that made up his medical interests. But when Covid struck, and Sam called them to say he would be starting work at Bellevue to help with the hospital's surge, they wondered aloud whether he could be taking it too far.

"I mean, sure, this is heroic, but I'm not really in favor of it,"

Neil said. "In a war you have some idea of where the bullets are coming from. But here you don't know. You have no idea where your enemy is. It's everywhere, really."

Sam's grandma, ninety years old and known for her dry wit, was the most opposed. (This was her boilerplate response to most of Sam's adventurous ideas—volunteering in Ecuador or Israel. She would quickly revert to enthusiasm once plane tickets were booked and plans cemented.) "Absolutely not," she told the family. "Bad idea."

But they all knew that Sam would make up his own mind.

Among Bellevue doctors, the first formal mention of the coronavirus came on January 24, in the weekly staff email: This is a novel coronavirus with its epicenter in Wuhan, China, with a ~14-day incubation period and potential for person-to-person transmission (though limited). . . . Risk to New Yorkers is considered low, and there have been no known health care worker infections.

Fast-forward seven weeks. As the virus began spreading across American states, shuttled on airplanes, transmitted at bars and bachelorette parties, the uneasy question of what space there was for medical students in this crisis hung over the hospitals.

The fourth-year medical students were untried. They had worked in the hospitals during rotations, but only under heavy supervision. Some school administrators felt a protective instinct toward these rookies. At the same time, they were also only three months away from beginning their careers, and it was evident that the hospitals would soon be overwhelmed, especially in New York. There was barely time to wrestle with these choices anyway. Crisis warps the speed at which time passes, every second both fleeting and costly. The siege was coming.

At first, on March 17, came an advisory from the Association of American Medical Colleges (AAMC), the main research and advocacy group for medical schools and teaching hospitals. The group recommended pulling all students out of their normal clinical rotations until at least the end of March. That would give medical schools time to develop safety protocols and learn more about this novel coronavirus. It would also allow hospitals to conserve protective equipment, which was in alarmingly short supply.

Fourth-year medical students around the country issued a rallying cry. They wanted to be useful. Ejected from clinical settings, they turned to other tasks. NYU students held a drive collecting surgical masks (8,200 of them) and N95s (700), plucking the rare goods from veterinarians, dry cleaners, nail salons, and tattoo parlors. University of Colorado students deployed to staff Covid call centers. A group of medical students at the University of Minnesota set up the MN CovidSitters so they could babysit for the children of overworked doctors.

But across the Atlantic, as the virus devastated Italy, European governments turned to more extreme measures. In March, Italy announced that it would send its final-year medical students into hospitals months ahead of schedule. British medical schools like Lancaster, Newcastle, and the University of East Anglia also accelerated graduation.

As New York's hospitals continued to brace themselves, city medical school deans began to reexamine their role in the fight. American hospitals had prepared for "discrete disasters" like mass shootings and hurricanes, but not for sustained periods of crisis like a pandemic. It was evident that staff would be overwhelmed. Texts went back and forth between students wondering whether they would soon be following in the footsteps of their European counterparts. There was no recent precedent for this kind of expedited move to hospital front lines. Medical school upperclassmen had been abruptly given the responsibilities of full physicians during the 1918 Spanish flu. Decades later, during World War II, schools created a fast-track program at the government's behest while physician shortages mounted. But since then, the rules governing medical education had grown more rigid.

Dr. Steven Abramson, vice dean for academic affairs at NYU medical school, kept receiving emails from fourth-years who said they wanted to help out. After all, these students had been trained by the medical workers now under siege at Bellevue and Tisch. Mulling this over, Dr. Abramson realized that these students had already completed all the curricular requirements—why not take them up on their offer? NYU sent a survey to fourth-year students asking if they would be willing to start work in the hospitals early. These temporary assignments would last just a few weeks in April and May. Within twelve hours, more than half had replied with a resounding yes. Soon after that, the process to get state approval for NYU's early graduation began. Other schools followed: Albert Einstein College of Medicine, Mount Sinai, Columbia.

Sam's graduation was set for April 3. On the allotted day, Jeremy slipped outside to pick up champagne. Sam opened his laptop and pulled up Webex. It was more than a month before their intended graduation. The virtual ceremony opened with stilted speeches from deans and faculty. Dr. Abramson greeted the group as "members of what we now call the COVID army."

Sam's laptop screen glowed with the familiar faces of his classmates, the fifty-two of them who had decided to graduate early. Some were wearing their caps and gowns, and one had made a diploma out of a manila envelope. Rows of Sam's friends dissolved into giggles as they stumbled through the technical difficulties of a graduation on Webex. Their voices bumped and tumbled over one another. No one was synchronized.

"I do solemnly swear, by whatever I hold most sacred—"

"I do solemnly swear, by—" "that I will be loyal to the profession of medicine and just and generous to its member"

"that I will be loyal to the profession—" "That I will lead my life and practice my art in uprightness and honor."

They paused for laughter. They scanned their eyes over the grid of classmates in bedrooms and family basements, all ensconced in those last moments of preprofessionalism.

"That into whatsoever home I shall enter it shall be for the good of the sick and the well to the utmost of my power and that I will hold myself aloof from wrong and from corruption and from tempting others to vice."

They couldn't squeeze each other's hands like in a normal graduation; there was for Sam no formal feel of a cap on your head, no thrill of a high-ceilinged auditorium. There was just his Greenwich Village apartment.

"That I will exercise my art solely for the cure of my patients and the prevention of disease and will give no drugs and perform no operation for a criminal purpose and far less suggest such a thing."

There were Sam's classmates reading too quickly, speeding through the final words of the oath as others stepped gingerly through. There were those reciting it steadily, like a heartbeat.

"That whatsoever I shall see or hear of the lives of men and women which is not fitting to be spoke, I will keep inviolably secret."

There was the weight of the words—the way you heard the pulse of each line when you were the only one in the room reading it aloud. There was the mess of sounds, the beeps and dings of fifty-two laptops, a harmony of machines not unlike the din of the hospital floor.

"These things I do promise and in proportion as I am faithful to this oath, may happiness and good repute be ever mine, the opposite if I shall be forsworn."

Sam heard Jeremy's cheers, and those of his classmates coming from his laptop. That night, every New Yorker's cell phone got an emergency push alert asking licensed health care workers to join the front lines and "support health care facilities in need." Sam's phone lit up with the message: "Attention all health care workers." Well, he thought, I'm heeding the call. He'd become a doctor that afternoon in his apartment.

Two

GABRIELA, Summer, mid-2000s

What Gabriela liked about working in her mother's salon was the cacophony of it all—the raucous chatter, the rush of water from the sinks. Gabriela worked as the shampoo girl. That meant she had to lather her hands in shampoo, massage into the scalp, rinse it out. Then again: lather, massage, rinse, repeat. The business was called Willow Salon and Day Spa. It used to bear Gabriela's family name, but no one could pronounce the Spanish properly in the lily-white town of Millis. Gabriela started shampooing there when she was fourteen, the youngest age she could legally work in Massachusetts, but even before then she liked to accompany her mom and spend afternoons offering $5 foot rubs.

Sometimes the clients, their heads tilted back, asked Gabriela questions as she rinsed: Where was she in school? What was she studying? What did she plan to do when she left home? Gabriela was asked so often that she had to form answers. "I'm going to be a doctor," she'd tell them proudly.

She would imagine, sometimes, what being a doctor would actually mean. Maybe it had something in common with being a shampoo girl, at least in the trust you had to earn. She'd run through some set of clinical tasks—check the pulse, check the blood pressure, rinse, repeat—then chat with her patients, ask them where they grew up and what they liked to do. She'd watch their smiles widen, seeing a girl like her, a brown face in a white coat. It didn't seem like an impossibility, or even all that distant. When Gabriela was younger, her mom insisted on driving her forty-five minutes out of town to see the only Hispanic pediatrician in the area, Dr. Vivez, over in Newton.

The salon's clients smiled when they heard Gabriela talking about her plans. "Your mom must be so proud," they'd say. Some were dubious: "Don't get your hopes up," they told Gabriela. "It's very hard to get into medical school."

But Gabriela's grandma, Grammy Beverly, kept her gaze dead serious. "Good for you boo," she'd say. "Keep your eyes on the goal."

Grammy moved in with the family when Gabriela was twelve. Gabriela's favorite place in the house was the kitchen counter where she could watch Grammy cook. Quietly, at first, just the sound of seasoned meat sizzling in a frying pan, but then Grammy would start to tell stories from work.

Grammy worked as a labor and delivery nurse at Baystate Medical Center, on Chestnut Street just off I-91. Most of her patients were teenagers; there were a lot of young mothers in the area. When they gave birth, Grammy said, their faces twisted in pain and fear and maybe ecstasy, and Grammy would hold their clammy hands and wipe their sweaty foreheads dry.

Grammy had an adage she liked to share with Gabriela, which felt sappy and poignant all at once: "For you it's a normal day. For the patient, it's either the best or worst day of their life. And you're in the middle of it. That's medicine."

Gabriela moved to New York the summer after she graduated from Amherst College. She took a job as a front-desk receptionist at a physical therapy clinic. Her new home was a fifth-floor walk-up on the corner of Sixty-First Street and First Avenue, an apartment so puny that her clothing closet was in the kitchen and her bedroom fit nothing but a full-size mattress. She was broke and far from family. But life still had that quality of omnipresent opportunity, as if any moment might sprout an adventure or career-altering event. Any chance encounter, on the subway or in the aisles of Whole Foods, could be the opening to an early-aughts Katherine Heigl rom-com. That was the aura of post-college life in Manhattan, each day crackling with maybes and what-ifs.

She matched with Jorge on Bumble that autumn. Right away she thought he was cute. Jorge was tall and broad-shouldered, with velvety eyes peering out from behind a set of thick-rimmed glasses. She swiped right and they messaged back and forth, mostly small talk peppered with Jorge's quips.

"Hey, this is going to sound weird for a first date," he wrote to her one evening. "But I got last-minute floor tickets to see the Weeknd tomorrow night. Are you free?"

Gabriela tried to force herself to slow her response, but she was too excited: "Yes!" The next day, stomach tossing, she took the subway to Atlantic Avenue, and they met outside the Barclay's Center in downtown Brooklyn. The arena pulsed, bodies illuminated by strobe lights flashing neon purple and red. The night had that quality of spending time with someone Gabriela had known a long time, even though she was with nineteen thousand sweaty strangers and a boy she had met only hours earlier. After the concert, she pulled up Bumble and canceled the date she had lined up for the next night. Months later she'd find out that Jorge had done the same.

For their second date, they met in Washington Square Park and played with his roommate's French bulldog. Right away their relationship took on the comfort and acceleration of something meant to be. They talked about family and their aspirations for New York careers. Jorge told her about moving from Bolivia to the United States when he was young and watching his parents learn English by working at halfway houses and rehabilitation programs. Gabriela told him about her own parents—her dad had moved from Ecuador to the housing projects of Holyoke, Massachusetts, where he met her mom, who was raised in Springfield's inner city.

Jorge met Gabriela's mom just one month after their first date, on a touristy December afternoon in Bryant Park, but the real reception that seemed to terrify him was Grammy. Gabriela warned him that Grammy had met some of her previous boyfriends and was always candid in her criticisms. She assigned enormous weight to her mom's and Grammy's judgments, because she knew the hell they'd gone through in separating from their own partners.

"Grammy's a very honest person," Gabriela told Jorge, as they were headed home to Massachusetts.

Gabriela's whole family gathered to greet them. She lived in a matriarchy of sorts, surrounded at all times by her mom, aunts, and cousins. Jorge went down the processional row. Grammy sized him up, narrowing her eyes. Her features were delicate, but she was somehow imposing, with her plump movie-star lips and mysteriously unwrinkled face. She had this exquisite blowout, not a strand of hair out of place. Jorge felt like he was greeting a dignitary. When Grammy announced her approval, Gabriela felt her veins flood with instant relief.

Like an empress, Grammy oversaw the goings-on of Gabriela's home with cool, keen judgment. Everyone wanted her approval, and she doled it out in the smallest of doses. "What are you doing on the couch?" she'd say. There were always more laundry loads to do, or plates to scrub. Grammy had been more strict than sympathetic when Gabriela came home in high school tired from her AP classes and exams. "You're lucky you have an education," she said. "Lots of girls don't have that."

Grammy knew what real hardship looked like, and she didn't have patience for petty complaints. She'd been raised by a Lebanese Christian father, who fled religious persecution and resettled in Massachusetts, where Grammy had raised her own children and started her hospital career. But anytime Gabriela got sick, Grammy dropped her stern tone and turned doting, bringing her granddaughter heaping plates of home-cooked food. Her specialties were kibbeh and stuffed grape leaves (which were labor-intensive: you had to soak the grape leaves, crush the pine nuts, then roll up the wrappings and tuck in the ground-beef filling like you were making little dumplings).

Gabriela knew that when she became a doctor, her bond with Grammy would only grow tighter. The two of them formed the professional branch of the family, those with access to that class of American life defined just as much by prestige as by a do-gooder bent. It was something they shared that ran deeper than their dark hair and arched eyebrows, more intimate than a genetic code. When Gabriela's mom wasn't certain medical school was a good idea, with all its years of arduous training—what about a nurse practitioner or a physician's assistant?—Grammy said: "Honey, she can do it. This is what she wants."

Gabriela took the MCAT during her senior year of college, two days after turning in her biology thesis. She filled out her med school applications that summer after college graduation, as she prepared for her move to New York. She could picture herself calling her grandmother up after a long day rotating at the hospital, sharing what she'd learned from her resident physicians. Grammy always reminded her not to stick up her nose at the nurses ("They can be your best friend or worst enemy"), and to remember that the science was important but the patient was most important of all. Grammy was retired from nursing at this point, mostly focused on helping to raise Gabriela's younger cousin who lived with them, so she liked the idea that she could pass on her decades of clinical knowledge to her granddaughter.

When Gabriela called her mom to say she'd submitted her applications, her mom passed the phone over to Grammy, who started crying. "I'm so proud," Grammy said. "I always knew you'd be a doctor."

When Gabriela started getting her med school acceptances that spring, she told her mom and Grammy that she'd go to Boston University so she could visit them on weekends. But then she got her NYU acceptance. New York had something that Boston didn't: Jorge. Then there was the magical pull of the city, all the hole-in-the-wall restaurants they'd just begun to explore. Gabriela decided she would stay in Manhattan, a verdict that her mom and Grammy begrudgingly accepted when she promised them frequent visits.

During Gabriela's first semester of med school at NYU, Grammy started showing signs of vascular dementia. She had suffered several strokes in her fifties, and it turned out that she'd had more recent ones as well that went undetected. She was diagnosed that fall with polycythemia vera, a type of blood cancer that thickened the blood and made it prone to clotting. With a few weeks of classes behind her, the clinical terminology didn't sound so foreign to Gabriela. But she wished that her first real-live test case, beyond the slides of her class lectures, could be anyone but Grammy.

Gabriela was only 170 miles from home, but after all the years she had spent living with her mom and Grammy, it didn't feel right that she shouldn't be there as Grammy's health began to deteriorate. Whenever she hung up the phone with them, she felt like she had to negotiate the terms for her mental return to schoolwork. Her classes were important too. It just didn't feel that way, with Grammy getting sicker. Soon Grammy no longer recognized her family members. In person, she mistakenly called Gabriela Val, the name of her daughter who had passed away decades before. On the phone, she called her Misty or Cookie, the names of her cats.

Grammy's health went into sharper decline during Gabriela's third year, while she was on a neurology rotation. At work, nearly all of Gabriela's patients had dementia. Back home in the evenings, she would call home for updates on Grammy and cry.

Gabriela and Jorge started making weekly trips up to Massachusetts. In April, Gabriela's mom pulled Jorge aside. "Hey, listen," she told him. "I don't know how much longer Grammy is going to be with us. And, you know, it was always important to Gabriela that Grammy see her get engaged."

The next weekend, when they drove up to visit Grammy in the hospital, Jorge proposed. For months he had been racking his brain for just the right setting for the proposal—bluebird sky, maybe a hired photographer obscured from view. But hospital visits put everything in perspective, vanities in particular. When they stopped at Grammy's hospital bed to show her the ring, she said her first clear words in months.

"It's beautiful," she murmured, looking up at the two of them. "I'm so happy for you."

It was the last time they talked with her. They drove back to New York that evening, so Gabriela could get back to her med school rotations, and Grammy died a few days later.

There is a photo of Gabriela, at around three years old, gently placing Band-Aids on her Big Bird stuffed animal. Throughout childhood, she never wavered in her medical ambitions. Yet all the same she encountered people dubious of her professional plans. And she wondered, sometimes, why it was that her mother had to drive her forty-five minutes out of town just to find a pediatrician who was Hispanic.

The profession remains predominantly white. Less than 6 percent of American physicians are Hispanic. Only 5 percent are Black. And roughly two-thirds of all doctors are male. Around half of medical students have fathers with graduate degrees, compared to 12 percent of men in the general American population. There's less evidence of just how many med students have parents who are themselves doctors, but the number is high.

The homogeneity of American medicine isn't a bug, but a feature of the system. Its whitewashed, wealth-washed nature was never accidental. In the late eighteenth and early nineteenth centuries, America's medical field was growing rapidly but without much formal structure, as Paul Starr elucidates in his sweeping history of the field. In England, the Royal College of Physicians touted a rigorous standard of professionalism and admitted only Oxford and Cambridge graduates. But on the wide-open American frontier, becoming a doctor was far simpler. Medical schools were opening across the country, especially in rural areas where costs could be kept low. By 1850 there were forty-two medical schools in America, and just three in France.

Mostly, these schools were roughly assembled with little oversight. A group of physicians approached a college with a proposal, and the college gave them the legitimacy to confer medical degrees. The instructors were unsalaried and got their compensation from student fees. Examinations weren't rigorous because students only paid their professors if they passed. The study terms were just three or four months. Getting a degree meant just two terms of study, a thesis, evidence of knowledge in Latin and philosophy, and three years of apprenticeship. But none of these requirements were too strictly enforced, because if any school was too rigorous, students would just go elsewhere for their studies.

Medical societies tried to impose harsher standards on the profession. The Massachusetts Medical Society, incorporated in 1781, was established because "a just discrimination should be made between such as are duly educated, and properly qualified for the duties of their profession, and those who may ignorantly and wickedly administer medicine." The group and others like it looked to the restrictive model of the Royal College of Physicians. But these societies didn't hold much sway in the United

States, because doctors could legally practice without their blessing. The number of schools and doctors in the country continued to multiply.

Early American medicine owed its loose norms partly to the country's lofty egalitarian aims. As Starr writes, the country's medical field "took its unique character in America from the dialectic between professionalism and the nation's democratic culture." But in the late nineteenth century, that democratic openness was called into question.

It started, unsurprisingly, at Harvard and Johns Hopkins, where the university presidents Charles William Eliot and Daniel Coit Gilman decided that medical schools represented the worst of the country's education system and needed reform. When Eliot became president of Harvard in 1869, he began personally presiding over medical faculty meetings. He was set on a cultural makeover for the profession. "The ignorance and general incompetency of the average graduate of American Medical Schools, at the time when he receives the degree which turns him loose upon the community, is something horrible to contemplate," Eliot wrote. He decided to redo curricular requirements: the academic term went from four months to nine; the minimum training period, from two years to three; lab work replaced some lectures. For the first time, students had to pass all their courses to graduate. Johns Hopkins similarly transformed its demands of its medical students.

At first those reforms caused a drop in enrollment, from 330 Harvard medical students in 1869 to 170 in 1872. But then the number began to climb again, up to 263 in 1879, likely a result of the new selectivity, and other schools, including the University of Pennsylvania, followed Harvard's suit. By 1890 the more advanced medical schools, around a third of those in the country, banded together and formed a national association, now known as the Association of American Medical Colleges. The AAMC set minimum standards for its member schools, including three years of training and required lab work.

At the turn of the century the American Medical Association, founded a few decades earlier in 1847, decided to make medical school improvement its own priority. It created a council on medical education and inspected 160 schools across the country, deeming 46 of them subpar and 32 unsalvageable. The group then invited the Carnegie Foundation for the Advancement of Teaching to assess the country's schools and produce their own recommendations for reform. The foundation appointed a young educator with a Johns Hopkins degree named Abraham Flexner. Flexner visited every medical school in the US and published his first report, *Bulletin Number Four: Medical Education in the United States and Canada* (best known as the Flexner Report) in 1910. Its conclusions were damning. Flexner thought the country had some of the best medical schools—but more of the worst. Many still had no labs, no books, and waived admissions requirements for anyone who would pay. As Starr described it, there were too many inferior schools and unskilled clinicians: "America was oversupplied with badly trained practitioners; it could do with fewer but better doctors."

The Flexner Report hastened the demise of poor and proprietary medical schools that were already in financial trouble. The number of medical schools in the country went into steep decline—and with it, the number of doctors. By 1915 America's medical school tally dropped from 131 to 95, and its graduates from 5,440 to 3,536. Seven years later those numbers were 81 and 2,529, respectively. The culling of American medicine had begun. Licensing boards then began to demand that those applying to be doctors have some years of experience in college as well; by 1922, thirty-eight states included this requirement.

The schools that remained in the years after the Flexner Report were mostly elite institutions. Many that shut down were designed solely for profit, and didn't provide adequate medical training. They waived their admissions requirements for any student who would pay; they didn't clean their premises properly, so the rooms reeked of cadavers; even their library shelves were empty. But the schools with fewer resources targeted for closure also included those in rural areas, those that trained Black students. In fact, after the Flexner reforms, only two of the seven Black medical colleges were left. A recent study in the *Journal of the American Medical Association* investigated how many additional Black physicians would have been trained as of 2019 had five of the shuttered Black medical schools been saved. Their estimate: as many as 35,315.

Even at the time, the Flexner reforms were criticized for the way they eliminated medical schools that supplied poor communities with doctors and gave poor people the chance to study medicine. At one such school in Tennessee, a doctor wrote: "True, our entrance requirements are not the same as those of the University of Pennsylvania. . . . Yet we prepare worthy, ambitious men who have striven hard with small opportunities." He ends with a plea: "Can the wealthy who are in a minority say to the poor majority, you shall not have a doctor?" As rural schools shuttered, they effectively did.

Meanwhile, the average doctor's salary multiplied by six between 1900 and 1928, largely a result of the field's rising professionalism and exclusivity. If no one was rejected from medical school before 1900, by the 1930s just half of applicants were getting in. In the wake of World War II, demand for physicians kept growing, so medical schools increased their class sizes. But the number of people applying rose too, and the field grew ever more competitive. Today, some 53,000 students apply for medical school each year, and roughly 40 percent are admitted.

What does all this mean in the context of a person like Gabriela at NYU? Students from underrepresented backgrounds enter a field whose training programs were designed for exclusivity. For over a century, medical education was structured to winnow down trainees and ensure that those licensed to practice represented the most educated, the best trained, those with proper pedigree. This is a sound aim: these are the people tasked with operating and treating our broken bodies, presiding over our health. But this aim has also made medicine a particularly challenging field to break into for those without access to high-quality early education and financial resources. It also means medical students and trainees don't always get the opportunity to learn from those who are different from them—people whose parents didn't go to college, those who grew up without health insurance or medical literacy.

For Gabriela, raised with the knowledge that few doctors looked like her, the road might have seemed unwelcoming. Yet she also knew, with all the assurance she felt washing hair in her mother's salon, that this was what she wanted: to make people feel at home in their own bodies.

Back at her NYU clerkship, in the days after her grandma died, Gabriela had to finish her neurology unit. Each time one of her patients woke up in a state of confusion or mistook the night nurse for his wife, Gabriela was awash in a fresh state of grief. She stumbled through the weeks of rotation with a sorrow that felt like her whole world had been draped in blackout curtains.

She decided that term that she would go into pediatrics. In a pediatrics department she wouldn't see her grandma in every aging face, bodies worn and memories failing. Besides, it felt like a specialty that still honored Grammy, in some way. For Grammy, medicine was all about the relationships and the comfort you could offer. It was about earning the patients' trust, which was essential for a pediatrician too.

Later that spring, during a pediatric dermatology rotation, Gabriela assisted an older physician in freezing the warts off a young girl's hands. The patient was dark-haired, around six years old, wearing a dress with an image of Princess Elsa from Disney's *Frozen*. She was frightened. She wasn't making eye contact with the doctor. Her mouth formed a horrified O when Gabriela came close with the liquid nitrogen gun. Gabriela's instinct might have been to grab her hand and pep-talk the child. But she watched as the doctor took a different approach.

"Want to see a magic trick?" he asked.

The girl nodded, cautiously optimistic about the offer. The physician took out a Dixie cup and poured some liquid nitrogen on it, then set it on the ground. The young patient's eyes widened as billows of smoke began to emerge. As she fixed her gaze on the reaction, mouth wide in marvel, the doctor bent down and applied the chemical substance to her hands to begin the freezing process. She didn't notice.

Gabriela and the pediatrician made eye contact then, and he shot her a sly smile. This was a skill of a different kind, not the sort assessed on medical school exams, and Gabriela knew it was the type of test that she could ace.

Some of Gabriela's classmates liked the tactile craft of medicine. They liked the dissections, the feel of a 25-gauge needle, tracing the curvature of a bone. Some liked the intellectual puzzle of diagnosis. Gabriela, buoyant and chatty and quick to earn trust, liked the connective tissue of a patient exchange. Like Grammy always said, the science mattered, but the patients mattered to her most.

Three

IRIS, February 2020

The year after she graduated from college, Iris fell in love with a boy from France. He'd been raised in the southern countryside, near Toulouse. The two met in a frigid suburb of Chicago in February, where they were both attending a retreat for first-years at the international consulting firm they had just started working for. Their connection was instant and easy. Benjamin had doll-like features set in a porcelain complexion, and Iris was a whir of kinetic energy. She was affectionate and tended to inspire immediate, fierce allegiance with her maternal airs.

When Benjamin flew back to France after the retreat, they kept in touch by WhatsApp, buoyed by the magical irrationality of long distance—the flood of early-morning texts Iris awoke to when it was already midday in Europe, the messages that stretched late into Benjamin's night.

Before the retreat, Iris had been weighing whether she should apply to medical school. She had studied to be a veterinarian in college, then abruptly switched to pre-med in her senior year. She had hoped that her time in consulting might lure her onto the sweet, smooth glide toward a corporate career. But instead she kept finding herself wondering about the work that existed beyond spreadsheets and conference calls, so she submitted her med school applications. She got into University of Southern California and started daydreaming about strolling on palm-tree-lined avenues, ditching the layers necessitated by Northeast winters. Then Albert Einstein College of Medicine called with a scholarship. The news was reassuring and restricting all at once. With the extra money on offer, Iris would have to stay in New York. By July, Benjamin had also secured a job in the city and a one-way ticket to the United States. They were married the next year, without any engagement (Iris didn't see the point).

Benjamin spoke near-perfect English; he worked for a large American company, as a supply chain consultant for medical devices. Still, Iris was set on mastering her French, partly for his sake and partly because she knew that if she tried hard enough, she could. But each time she practiced during their visits to his family, she wound up ashamed of her accent and relying on Benjamin to translate. She could literally see the French words floating in her mind, even inching toward her tongue, but she couldn't formulate them. It wasn't that she was excessively proud, but she didn't want to offend people by butchering words.

So Iris decided she would vie for one of Einstein's coveted study-abroad spots in her fourth year, and complete a rotation at a hospital in Paris. There she could practice speaking far from Benjamin and his family, without any concern of high-stakes appraisal. She was ready to come back rolling her *r*'s like it was *rien*.

Benjamin drove her to the airport in mid-February. On the way, he asked her about the virus that had devastated Wuhan. Benjamin read French newspapers like *Le Monde*, and reporters there seemed concerned about the disease making its way across borders, transmitted through international flights like the one Iris was embarking on. Iris said she hadn't been following it much. Neither of them realized that the virus might have already reached New York, circulating in supermarkets and doctor's offices long before the governor and mayor took action to shut the city down. They didn't know of reports that would suggest that the first coronavirus patient in France, a coughing fishmonger, had actually turned up in a Paris hospital on December 27.

Some of Iris's classmates had briefly discussed the stories from Wuhan two weeks before, when the World Health Organization declared the outbreak a global health emergency. But Iris wasn't too worried. She'd studied coronaviruses in her first year of medical school; usually they caused a common cold. Sometimes they could lead to severe acute respiratory syndrome (SARS) or Middle East respiratory syndrome (MERS), but typically they were benign. And anyway, Iris was more focused on crepes and coffee shops, on weeks of ambling through the eleventh arrondissement before she came back to celebrate graduation.

For her first days in Paris, Iris forced herself to blissfully tune out the news. She had two weeks of language study before starting in the pulmonology department at the hospital. She spent her afternoons memorizing terms like *la bronchite* and *le pneumothorax*. At night she wandered the cobblestone streets and people-watched in bustling bars.

One evening her dad called from Shanghai, where he often spent time for his work. "Are you being careful?" he asked. "Make sure you're wearing a mask and covering your eyes in the hospital. This virus is no joke." He told her about how the whole city had shut down around him. Businesses had shuttered, and towering office buildings emptied out. There were frightening reports of young people turning up in emergency rooms unable to breathe, and in some cases dying soon after.

But this all seemed far away. Iris was so focused on her language studies that it wasn't hard to redirect her attention. Some of the diseases she was seeing in the Paris hospital were rare in the United States, the type of cases she had only read about in textbooks. In just a week of clinic she'd seen more of the inflammatory condition Behçet's syndrome, also called Silk Road disease, than she'd thought she might see in her entire career. It was more common in Arab populations, and there was a large community of Arab immigrants near the hospital. She saw idiopathic pulmonary fibrosis and chronic obstructive pulmonary disease.

During her second week in the hospital, a patient showed up wheezing and feverish. He was whisked away by doctors in full protective gear. Everybody on Iris's floor started whispering about whether that patient had the novel disease.

Back in the States, Benjamin was getting more worried too. There were reports of the virus circulating in Seattle. There were Americans trapped on cruise ships with sick passengers abroad. The markets were tanking.

On March 11, wiped out after a long day on the wards, Iris went to bed early. That night, while she slept with her data turned off, Donald Trump got on television to suspend American travel from Europe. Iris woke up to a barrage of frantic texts. Her friends wanted to know: Was she stranded in Paris? Benjamin had written her to say he'd purchased her a ticket home.

The next day was a blur of packing and goodbyes. Over the holidays, Iris's in-laws had bought her a leather-bound French-English medical dictionary, the *Dorland dictionnaire médical bilingue français-anglais*. Iris had liked it so much that she'd ordered another copy to give her attending physician as a parting gift. But it didn't come in time for her abbreviated departure, so she went to the hospital Thursday morning and gave him her own copy. She spent that afternoon furiously shoving her belongings into two suitcases. The night before she left, she went for drinks with a friend, who said she had never seen the Paris streets so empty.

Iris's flight from Charles de Gaulle was stuffed with Americans trying to get home before the borders closed. Plenty of people, Iris and her seatmate included, weren't wearing masks—the surgeon general had recently told the public to stop hoarding face coverings and save them for frontline physicians. But most people carried Clorox wipes or little bottles of sanitizer, and Iris watched as her fellow passengers sanitized and scrubbed their seats maniacally, as if they were synced up in some cultish rite. There was just one empty seat, which Iris was permitted to switch to after she knocked over a cup of water and soaked her own.

Iris had a thousand questions about life on the other side of the trip—Would she and her classmates be kicked out of the hospitals? Was New York City really shutting down?—but she was exhausted from her last day of work and packing, so she slept the whole way home.

When she landed in a swarmed New York airport on Friday, March 13, nobody took Iris's temperature, or asked her if she was feeling ill.

On the customs line, she turned her data back on and pulled up her email. She composed a note to Einstein administrators, telling them she would be happy to help out at Montefiore Medical Center if they were considering bringing in medical students. She added that she had plenty of clinical experience, including from her recent weeks in Paris.

Iris felt this distinct sense of duty when she thought about patients rushing the city hospitals. The weight of the crisis was beginning to feel real, and she realized that she was in a unique position. Unlike Benjamin and her family members and even the newscasters watching the outbreak spread in helpless horror, Iris had been trained with the skills to help out. She was supposed to be starting a residency in internal medicine at Montefiore in three months anyway.

The school dean responded within a few minutes to say they appreciated Iris's offer, but all students had just been pulled out of rotations.

Iris sighed, looking around at people shouting and shoving their way toward the doors. There were points of possible transmission all around her: hands brushing at baggage claim, children sprawled on the linoleum ground, carts slung from passenger to passenger. She knew she was on the precipice of a historical moment, but with weeks until her graduation, she also felt like just a liability.

When Iris was growing up, no one could persuade her to drink milk. It wasn't a dairy moratorium in general; she didn't mind ice cream. But there was something about milk, its raw creaminess, that made her gag. Which is not particularly important, except to explain why it wasn't a surprise to anyone in her family that her bones grew in brittle. They attributed it to a lack of vitamin D.

That Iris's bones were brittle was discovered when she was still a kid, on the playground near home in Flushing, Queens. It was a sticky day in July, one of those humid afternoons when people stick to the park sprinkler systems like they've got magnetic pull. Iris was playing a game with her little sister, the two of them hopping across a set of low hurdles. Then one of Iris's feet got tangled with the other, and she tipped forward. Right away she heard a crack. She looked toward her arm and saw her elbow popped out of place, her hand bent at a mechanically disconcerting angle.

Iris let out a howl. Her mom and grandmother sprinted across the park, each appearing at her side in seconds. Jolts of pain shot through Iris's body, and she continued to wail. Then her grandma warned her it was about to hurt even worse, and she pushed Iris's dislocated shoulder and elbow back into place.

When she looks back now, it's hard for Iris to remember exactly what happened in the following minutes. Partly it's because of the time gone by, partly that she probably never took in all the details in the first place. (The Ebbinghaus curve, which maps our tendency to forget, says both time and relative strength of a memory can dilute its details; in this case, over a decade has passed, and the moment has already been blurred by youth and by injury.) But Iris remembers her grandmother and mom exchanging words in hushed tones, as they considered whether or not she should go to the hospital.

Her family had an innate skepticism toward the American medical system. Iris's mother, raised in a village four hours from Shanghai, worked in traditional Chinese medicine, doing acupuncture and mixing herbal remedies out of a clinic in Manhattan. In their household only the most severe ailments were treated with something inorganic. A cough or sore throat merited a mysterious tea; it was only full-blown days of fever that led to the rare dose of NyQuil or Tylenol. So the idea that Iris would be carted into a Queens emergency room and turned over to white-coated, non-Chinese-speaking figures was unsettling.

As Iris's tears continued unabated, the two women decided they should seek treatment for the disturbingly crooked arm. They took her to the closest hospital. It was Iris's first time in an emergency room, and she tried to make herself as unobtrusive as possible, listening as her mom and grandmother argued in Shanghainese.

As they waited and the hours dragged on, one of them determined, for some reason, that the hospital might call the police on them. It's tough to say where this idea came from. It wasn't founded in anything, except the nebulous worries distinct to immigrant families that admission to any public American institution might entail some level of risk.

The question settled over their conversation like a storm cloud—Will we go to jail?—and the two women argued while Iris bawled. She watched the stream of midsummer weekend injuries flowing through the hospital's entrance, along with harried doctors in their brightly colored scrubs. The sky outside turned lilac and then navy as the sun disappeared.

When the doctor finally came over, Iris was confused by his genial smile. She'd come to expect someone foreboding just from the look on her mom's face. But he had the even tone of a first-grade teacher as he introduced himself. She remembered thinking that for all her family's panic, this place of strange uniforms and white walls wasn't all that bad.

It turned out her arm was broken. The friendly doctor fixed Iris up with a plaster cast, which made her arm feel heavy, detached. Iris stared at the blank canvas with the forlorn realization that because it was the middle of summer, she wouldn't be able to get all her classmates to cover it in signatures and doodles. Last year, everyone in Iris's elementary school with broken bones had gotten their casts signed by classmates with notes like "BFFs" or "I <3 You!" or "you rox my sox." All Iris got was a summer of lugging around her boring, extra heavy limb and wishing she could scratch at the skin underneath. She felt the sting of injustice.

Iris can still remember her mom's look of relief when they finally trailed out of the emergency room around midnight. It was like she hadn't let out a full exhale the whole time they were there. And Iris would wonder for years what it was about medical facilities that so terrified her family—whether it was the reams of paperwork they couldn't understand, or the serious faces that froze in confusion at their accents. When Iris was seventeen, her paternal grandmother died of renal cancer. The case probably could have been treated easily if it had been detected early, but she refused to take herself to a doctor until the cancer had progressed to stage four. By then it was too late. She was scared of English-speaking doctors; the only English phrase she could understand was "One dollar, one dollar," which she learned from the street hawkers selling wares by her home in Harlem.

There was something about hospitals—call it the Americanness—that made Iris's family feel like they didn't belong. This feeling would make them all the more proud when Iris was accepted to medical school at Einstein. It was hard to believe—the notion that their own daughter could not just understand but master the intricacies of the American health care system. But Iris had always aspired to fluency, not just in language but in American culture. When she spent summers back in Huangshan (the Yellow Mountain region) and Shanghai, visiting family members—

just one cousin on each side, a result of the country's one-child policy—she always felt distinctly out of place. She wished that she could be back in New York with friends, playing by the local pool.

If she couldn't fit in at her parents' place of birth, she would blend into her new home. And there was no path to belonging that was quite as direct as medical school was.

Throughout medical school, the most off-putting part of American medicine, to Iris, was the paternalism she saw in older, old-school physicians. There was often this assumption that the doctor knew best. That could be true, but it could also be a detriment to the diagnosis. And it meant that doctors could miss out on hearing the knowledge the patients brought outside of textbook anatomical terms. Her own mom, for example, couldn't tell you how her body might react to a benzodiazepine, but she had a nearly inexhaustible mental catalog of herbs that could cure common aches. That was her own type of medical literacy, one that didn't play by American rules. In busy US hospitals there wasn't always time to get to know the patient, only extract the necessary data.

The extent of this brokenness in the system hit Iris in the summer after her first year of med school, when she got diagnosed for depression. The diagnosis came through a rote recitation of symptoms by a psychiatrist: Have you had any changes in sleep patterns? Have you been depressed or down? Have you had any changes in appetite?

Yes, yes, yes. But the doctor quizzing Iris made her feel like she was running through the items on a grocery list. Like she was getting her prescription at the expense of her sense of wholeness and individuality, the feeling that these complex emotions were hers and hers alone.

The worst were the doctors who said "I know how you feel." These doctors probably didn't fully know how she felt, so what else were they getting wrong? But she stayed glued to her clinic chair, because she had the sense that she'd boarded a train that was now barreling along at breakneck speed and there was no way she could climb off.

At least American medicine had a vocabulary for Iris's kind of pain. Among Iris's Chinese relatives, "mental illness" meant people with dilated pupils and slurred speech. Not someone high-functioning who happened to have depression. There were no gradations. When Iris decided to make an appointment in the Montefiore psychiatric clinic, she didn't mention it to her parents; she'd only tell her mom a year later, when she was well into her medication course. She didn't feel her parents would know how to make sense of the agonizing feelings that overtook her sometimes, on the mornings when she didn't want to get out of bed. Iris's parents had survived China's cultural revolution; her own survival stories, meanwhile, were of finals week at Wellesley College. To call herself depressive seemed like nothing but a big whine.

On the American side of the medical system, when it came to depression, there was a know-it-all attitude so stifling it made Iris want to avoid doctors entirely; on the Chinese side, there was silence.

All of this weighed on Iris during her third year, when she rotated at Bronx Psych. She spent six weeks at the psychiatric hospital, a glassy building that curved inward like a parenthesis. She was paired with a single patient the whole time, a sixty-year-old man with schizophrenia. Bronx Psych was known for its large population of incarcerated patients. Some of Iris's classmates had patients who were hard-edged or aggressive, but Iris's patient, Mr. Nelson, seemed mostly concerned with her safety at the facility. If any of the surrounding patients began to yell or thrash, he'd turn to her: "Just don't mind them," he'd say.

Reading through Mr. Nelson's thirty-year file was like scanning a straight-D report card: only negative notes. You could almost hear the disapproving tut-tut-tut of whoever had previously filled it out. There were records of violent outbursts, hallucinations, delusions, confusion. Iris found herself adding comments on all of his best qualities. "He's remarkably resilient," she wrote. One day Mr. Nelson asked her to help him review the alphabet—he couldn't read or write—and they spent an hour carefully retracing each letter. She watched his brow furrow in concentration, and in her report she added: "He's eager to learn."

Some days, Iris wanted to tell him that she had been on his side of the medical equation.

"We're not so different!" she would say.

But then it felt like an unnecessary weight on their relationship. Maybe he didn't want the burden of that knowledge, which could be a bond but just as easily an intrusion.

When she was growing up, Iris's image of a doctor was of someone distant, commanding, and cold. She knew that wasn't the type of physician she wanted to be. But as she moved into the field, the norms that govern doctor-patient relationships started to shift.

Throughout much of the twentieth century, doctors often made decisions for their patients. They determined whether patients should be given a drug or operation. Patients weren't even permitted to see their medical records; the file wasn't considered their property. Some say the phrase "Give it to me straight, Doc" had its roots in this medical era; unless you asked explicitly, the doctor might not.

This distance between doctor and patient reaches back at least to the invention of the stethoscope in 1816. With that silver apparatus, doctors could extract information about the patient without actually pressing ear on chest. There was no skin to skin, just a device that allowed the physician to know something the patient couldn't. That sense of both authority and remove became essential to medical work; it helped transform what was once a trade into a profession. For more than a century, it remained sacrosanct.

That began to change in the mid-1900s, accelerated by the 1947 trial of twenty-three Nazi doctors and bureaucrats. They were indicted for torturous and unethical experimentation on their Jewish and other victims that included mass sterilizations, bone grafting, and forced exposure to drugs. The physicians claimed as part of their defense that no medical code of ethics limited their behavior. The Nuremberg Code that emerged as a result called for the "voluntary consent" of subjects in human research—in other words, a patient had to know what was being done to their body.

Still, giving consent was not the same as participating actively in medical decision-making. The rights of the patient gained further attention in 1984 when Dr. Jay Katz, a Yale doctor and ethicist, published *The Silent World of Doctor and Patient*, which criticized modes of clinical decision-making. Dr. Katz thought patients ought to be involved in their own medical choices. Patients should have a voice, and their relationship with providers ought to involve a two-way channel for trust. The book added new fuel to conversations about patient rights already underway in the field. Medical schools added it to their curricula and started to teach ethics classes about patient autonomy.

In part, the new emphasis on patient-doctor relationships was a revolt against decades of increasing corporatization in American medicine. For centuries, your local doctor might have been a next-door neighbor or a town healer. But throughout the mid-1900s American hospitals grew to be large, moneyed enterprises. By the 1980s medical workers seemed to patients less like their neighbors and more like the emissaries of sprawling corporate entities. Titles like "head nurse" changed to "clinical nurse-manager," a role that required expertise in compliance and the business of medicine. Dr. Bernard Lown, a physician and Nobel laureate, thought that the biggest problem in health care was not about economics or insurance but compassion. In a book that circulated widely among physicians, *The Lost Art of Healing*, he bemoaned the absence of doctor-patient bonds: "Healing is replaced with treating, caring is supplanted by managing." Instead of tending to a full human, physicians treated distinct organs, like a car mechanic examining malfunctioning parts. Doctors had learned how to operate, but not how to listen.

For some doctors, truly hearing a patient's voice is a skill that can prove more difficult to master than dressing a wound or reading an X-ray. Rana Awdish, a critical care physician in Detroit, discovered this firsthand when a mass in her liver ruptured during pregnancy, and she was admitted to her own hospital's ICU. Suddenly transported to the other side of the doctor-patient relationship, she realized how often doctors dismissed patient concerns in the course of their busy days, even when the patient was raising an issue essential to the diagnosis. "We aren't trained to see our patients," she wrote of the experience. "We are trained to see pathology."

But what exactly it should look like to respect a patient's wishes wasn't clear. Should the patient have complete independence and be free to dictate their treatments at will? Should the doctor seek to inform or even mold these decisions, according to what they thought was best? Or, through a more complex route, should the doctor seek to shape clinical choices according to the patient's personal values, which they had taken the time to learn?

Doctors Linda and Ezekiel Emanuel wrestled with these questions in a paper published in 1992, "Four Models of the Physician-Patient Relationship." The first model, the more old-school paternalistic type, the authors effectively dismissed. But the other three—informative, interpretive, and deliberative—they weighed carefully. The informative doctor provided factual information so that patients could make their own choices. (For example, they might say: "If you choose to be resuscitated in case of heart failure, your ribs could break.") The interpretive doctor supported patients in articulating personal values relative to their care. ("Let's say that after lung failure, you might never recover the ability to breathe on your own—would you still want to be resuscitated?") And the deliberative doctor went the furthest, helping patients to reexamine and possibly shift their personal values in light of the medical situation at hand. ("What parts of your independence are so valuable to you that a life without them wouldn't be worth living?") While the authors thought the informative doctor was most common, they saw the deliberative one as the ideal.

But as anyone who has ever been sick knows, patients don't always want to make their own choices. They come to doctors because doctors are the experts. It can be quite difficult to make a clear-eyed, rational decision about your own body, especially in a crisis. It's also difficult for family members to make a rational decision about Grandma, Mom, or Dad. *The Practice of Autonomy*, published in 1998 by Dr. Carl Schneider, a professor at the University of Michigan, brought new nuance to the conversation on patient rights. Dr. Schneider made the case that patients frequently don't want to be left to their own devices. A debilitatingly sick patient is likely thinking more about their nausea, pain, or exhaustion than the logic of a clinical decision.

"Where many ethicists go wrong is in promoting patient autonomy as a kind of ultimate value in medicine rather than recognizing it as one value among others," wrote Dr. Atul Gawande, a surgeon, writer, and Harvard medical professor, in a 1999 essay for the *New Yorker.* He described treating a patient who had just gone into sepsis, his immune system attacking his body as his bloodstream flooded with chemicals to fight off infection. A chief resident calmly told the patient he would be intubated, to which the man responded: "Don't . . . put me . . . on a . . . machine."

What were they to do? The man was young and otherwise healthy, with a wife and child. The procedure could easily save him, which his wife begged for the team to do. When the man fell into unconsciousness, the doctors put him on a ventilator.

His words when he awoke the next day were simply: "Thank you." Often doctors do know best.

Two decades later, conversation about the optimal relationship between doctor and patient continues to evolve. New generations of doctors are keenly aware of the ways that race, class, and culture affect this relationship. Dr. Uché Blackstock, founder of Advancing Health Equity, points out that white doctors tend to be more verbally dominant when speaking to Black patients. They sometimes carry biases that prevent them from taking their Black patients' concerns seriously. There is often a level of mistrust in the relationship shaped by histories of medical racism.

When patients distrust their doctors, or don't feel like their providers are truly listening to them, it can be challenging for them to articulate their values. They might struggle to have candid conversations about how to apply their values to medical choices. And they might waver, most of all, in trusting that their doctors have their best interests at heart. These are the aspects of medical care that have historically been harder to teach; in 2005, a study in the *Journal of the American Medical Association* found that resident physicians felt far more prepared in clinical and technical areas than in what they called "cross-cultural competency," communicating with people across social and cultural divides.

Physicians like Iris, Gabriela, and Sam see this as one reason that their identities should inform their work. They can connect with their patients on the basis of their personal histories—drawing on their multifaceted ethnic, sexual, and religious identities. They are pushing to humanize the field. Because doctors are not demigods; neither are they automatons. They are people, people who fall in love and doubt themselves and care passionately and sometimes make mistakes. They are people who themselves get sick and see family fall ill. That humanity makes them no worse at their job. It is in fact an essential quality. And nowhere was it experienced more viscerally than in the Covid wards.

Four

GABRIELA, March 2020

"Coronavirus Death Toll in China Crosses 3,000." "W.H.O. Declares a Global Emergency." "A Patient with Coronavirus Has Died in Washington State."

These didn't sound like Gabriela's reality. Yet there the words came, dispatches that turned from unthinkable to incontrovertible. Gabriela's mom followed the headlines and then transmitted her worries to the rest of the family with all the dutiful reluctance of a prophet come to foretell their fate (she watched MSNBC with religious devotion). Effectively, the message was that they were screwed.

By March, Gabriela's mom was starting to worry that she would have to temporarily close the salon. "What will my girls do?" she worried aloud to Gabriela, over the phone. Her eighteen workers called each other sisters and were in one another's weddings. Some of them had started working at the salon when they were teenagers, and now they were in their thirties with kids. The staff was family.

Growing up, Gabriela knew the series of noises that meant her mom was home from work: the front door opened, keys jangled in the kitchen—then the front door opened again, and the car restarted in the driveway. Her mom was always coming home to start dinner, then making one last trip back to the salon. Just in case a hair straightener was still on or a dryer out of place. She worked fifty hours in a good week, sixty in a normal one.

Gabriela could still remember when her mom decided to start the salon. She bought *Starting a Business for Dummies*. She pored over the book nightly, curled up in bed with the comforter bunching up under her toes. Gabriela could tell her mom was proud to be taking this step. She'd never gone to college, but she was going to be a business owner.

She found a spot she liked on Main Street in Millis, a town near their home in Medfield, then gutted the building so she could redesign it just the way she'd imagined. It was white-walled, with cherry oak rafter beams that gave the space the feel of a ski lodge. There was glamour too, with a multitiered chandelier dangling from the room's center. The little tables were covered in combs, hair spray, and vases of flowers. The seats were just five or so feet apart from one another so the clients and the stylists could chat. All day the ladies exchanged jokes over the roar of the dryers. Gabriela liked to watch her mom move through the salon, because it was where she seemed happiest. Calling out instructions, asking after someone's kids, looking around this space that she had created.

But as she watched the news of the virus making its sweep across Europe, Gabriela's mom knew it was only a matter of time. "We're going to have to shut down," she told Gabriela.

Meanwhile, Gabriela had read that final-year students in Britain and Italy were being given the chance to graduate early. Part of her wanted to feel useful when the virus hit New York. The other part of her was just afraid.

The email from NYU about working in the Covid wards came on a Tuesday night. Gabriela knew it was serious; why else would the school send out a notice to students at 7:00 p.m.? What was so important that it couldn't wait twelve hours for the next morning? The note started with discomfiting honesty: "We need your help."

Gabriela and Jorge were partway through a movie when the email came. She hit pause and opened it, reading silently. With the growing spread of COVID-19, our hospitals inundated with patients, and our colleagues on the front lines working extra-long hours, we are still short-staffed in emergency and internal medicine. Burnout of our doctors has become a growing concern.

The email went on to outline an opportunity for final-year students: they could graduate early and begin working as interns at NYU-affiliated hospitals as early as April, instead of July, when they were set to start residency.

"What's up?" Jorge asked.

"It sounds like they might be graduating us early." Gabriela stared at the screen, perplexed. "They're giving us the option to work in the hospitals and help with Covid."

Gabriela had an instinct, programmed by years of teary conversations, to dial her mom's number the moment she felt a certain stinging uncertainty. She picked up her phone to plug in the familiar digits, then stopped herself. She could already hear her mom's response, as though it were scripted: *But it's so dangerous.* On top of that, it seemed wrong to pile her worries on top of the family's existing stress about the salon.

Gabriela wished that she could call Grammy. Like a counter to her mom's script, she could hear what Grammy might say. She would listen patiently, then tell Gabriela that this was what the field was all about. This was why she had chosen to be a doctor.

She turned to Jorge. "Do you think I should do it? Would you feel safe?"

He gripped her hand. "If it's what you want to do, I wanna support you."

She and Jorge began to plan out routines for sanitizing right away, and she knew there was no doubling back. They would put a cardboard box at the apartment entrance for Gabriela's work items. They didn't have a washer-dryer in the apartment, but Gabriela would wear leggings under her hospital clothes and wash her scrubs in the shower the moment she got home (if she sent them out to a laundromat, she could leave someone else exposed to the virus). They would both be vigilant about washing their hands, zealously sanitizing, avoiding touching their faces.

Newspapers started reporting on NYU's early graduation the next day, so Gabriela knew she would have to call home about it. "Okay, don't get mad at me for not telling you this right away," she told her mom on the phone.

"You're not seriously thinking about doing this, are you?" her mom asked.

Gabriela realized that her unease was dissipating fast. She knew she wanted to go into the hospital. It wasn't that she didn't feel scared. She did. But her fear felt thin, like a worn layer covering something deeper and more substantial. She wanted to do what she'd been trained to do. She wanted to do what she knew would have made Grammy proud. It wasn't the culmination of medical school that Gabriela might have otherwise chosen, but it was hard to ignore that her training, and her aspirations born long before training, had prepared her for this exact moment.

Gabriela's Webex graduation was on April 3. The video captures her bursting out in laughter midway through the Hippocratic oath. In the background, a chorus of her classmates can be heard reciting the words of the oath, some solemn and others giggling like Gabriela. She is wearing her white coat and a formal dress underneath. Her screen is a study in contrasts: her hair is carefully styled, her face done up, but in the background is the goofy portrait of a llama hanging on the living room wall.

Back around the holidays, Gabriela's mom had booked a hotel room for graduation weekend. She planned to drive down from Massachusetts with her husband Tommy, Gabriela's stepdad. They would all go out to dinner and drink wine and giggle when Gabriela's mom got sentimental.

Instead, Gabriela was in her apartment, on the blue suede sofa with Jorge, who helped fashion a conical graduation cap from cardboard. Jorge had ventured out to buy champagne and charcuterie. When he left, he'd pulled on gloves and a face covering, like he was leaving on a covert military operation and not a trip to the liquor store. They weren't used to taking all these precautions for a simple errand. But this was what the pandemic had done: turned the mundane into some cross between heroism and paranoia.

As Gabriela listened to a speech from her dean, she felt a wave of grief on behalf of her family. Medical school had been a tiring, uncertain path. She and her mom deserved this moment, to make eye contact with each other across a grand auditorium. They deserved to squeal and take too many photos and wrap each other in suffocating hugs. Or they deserved to at least be physically together. But then, on FaceTime, Gabriela caught sight of her mom's face—eyes dewy, cheeks moist.

Later, on the phone, Gabriela's mom marveled that the emotions felt even more intense than they might have in person. "You know, that was probably more memorable than it would have been if it was just a typical graduation," she said. "It was very moving for me."

"Yeah," Gabriela agreed. "It had more meaning."

There is little in life as unshakable as the overconfidence of a New York politician. This is a fact that, among many others, played some role in the virus's catastrophic march through the city, as Gabriela waited for her call to the coronavirus wards.

"Excuse our arrogance as New Yorkers—I speak for the mayor also on this one—we think we have the best health care system on the planet right here in New York," Governor Andrew Cuomo said on March 2. "So when you're saying, what happened in other countries versus what happened here, we don't even think it's going to be as bad as it was in other countries."

It was worse.

A surge of patients hit the hospitals in late March and early April like a tsunami. At the city's first-wave peak, the hospitals were making room for nearly seventeen hundred new patients a day. Intensive care units were at capacity, hospital lecture halls filled with patient beds. Medical tents were erected in Central Park by an organization that had previously deployed to treat Ebola in the Democratic Republic of Congo. The USNS *Comfort*, a hospital ship, floated semi-uselessly in the Hudson. By mid-May, the city had seen nearly twenty thousand deaths. The media reported the crisis with breathless horror; Twitter and Facebook were flooded with videos of doctors and nurses pleading for spare protective equipment.

For families watching the spread of Covid-19 from far away, New York City seemed like the tenth circle of hell. The city and its surrounding suburbs became the epicenter of the pandemic in the United States, while the country became the epicenter of the pandemic in the world.

The city's density played a role in the disaster. So did its function as a hub of commerce and tourism. But New York's response was hampered too by confused guidance from the federal government, which minimized the virus's threat, as well as dysfunction at the state and local levels.

New York had a pandemic preparedness and response plan, which was drawn up in 2006 and ran hundreds of pages long. The report predicted that a flu-like sickness could infect large numbers of people and overwhelm the health care system. It anticipated two problems that the state would do well to address in advance: shortages of up-to-date emergency equipment, and a limited number of available hospital beds. It also highlighted the fact that health care workers would be at high risk of infection, which would add further pressure to the system.

But the papers were just that: papers. There were few concrete recommendations for what officials could do to prevent or allay the distress that the report so presciently described. As health care executive and Cuomo advisor Michael Dowling told ProPublica, "A plan on a piece of paper that doesn't have an operational part means nothing."

On March 1, the city announced its first confirmed case: a thirty-nine-year-old woman who had landed a week earlier in John F. Kennedy airport on Flight 701 from Doha, coming home from Iran, where Covid-19 was already running rampant. The state reported its second case, a lawyer from New Rochelle, the day after. Alarmingly, he hadn't traveled to any affected country. That confirmed everyone's worst suspicions: New York was already experiencing community spread. Though the lawyer had traipsed through crowded Manhattan that week, the state clung to its futile strategy of containment and tried to lock down New Rochelle.

It took three more weeks for New York, state and city, to begin shutdowns. If the city had implemented widespread social distancing measures a week or two earlier, it might have reduced its death toll by well over 50 percent, according to an estimate from Dr. Tom Frieden, former head of the Centers for Disease Control and Prevention. But both New York's governor and the city's mayor were also having to weigh what it would mean to shut down an economy as large as New York City's, and the panic that would no doubt ensue. Whether Mayor de Blasio was giving appropriately weighty consideration to every tough issue would later be challenged, as the city lost confidence in him. ("We'll tell you the second we think you should change your behavior," he cheerily intoned in early March.) But early on, even the head of New York City Health and Hospitals questioned the merits of a shutdown.

The city and state's leaders dug in their heels until March 20, when Cuomo finally announced what he called "the ultimate step," banning all nonessential gatherings of any size and mandating that nonessential businesses send their workers home. That was the Match Day when Gabriela, Sam, and their classmates popped champagne and FaceTimed family, each wondering what all of this would come to mean. The governor's executive order took effect two days later.

The state and city were also slow to address the crisis that the pandemic report foresaw at hospitals, including shortages of protective equipment and ventilators. It took Cuomo until mid-March to set up a task force charged with, among other things, increasing the number of available hospital beds. He directed hospitals to expand their capacity by at least 50 percent. Toward the end of the month he created a command center to track levels of emergency supplies and available beds, directing resources to the regions in greatest need.

To frontline providers, much of this looked like trying to stop a tidal wave with a sand fortress. America's hospitals operate on what *Atlantic* writer Ed Yong calls a "just-in-time economy." Instead of stockpiling masks, nasopharyngeal swabs, and painkillers, the hospitals acquire supplies as they're needed, in a byzantine chain that snakes around the globe. The material for face masks comes largely from Hubei, China; swabs are often manufactured in Lombardy, Italy; and many medications depend on production lines in China and India. When the pandemic hit, these supply chains broke down. Meanwhile, medical workers in the US fashioned masks from bandannas and scrambled to ration out swabs.

By summer, it was clear that some of the early expert predictions for the city hospitals hadn't come to pass. Fewer than 19,000 people were hospitalized by the April peak, while it had been projected that the number could hit 110,000. But there were still hospitals that ran out of beds, others making use of rooms that had long been dormant, and alarmingly low levels of supplies, including personal protective equipment.

So it was sensible that faraway relatives like Gabriela's mom would be afraid. The photos from Wuhan and Lombardy were devastating—half-naked victims lying facedown in a mess of wires, figures floating around them in hazmat suits—and it was assumed that New York City would be next. Some compared locked-down New York to London during the Blitz: the same dread, the silence broken by sirens, the sense that things, as bad as they were, were only going to get worse.

Five

JAY, September 2015

It can be hard to recall when a place began to feel familiar, and easier to remember your first arrival there, the feeling of being woefully out of place. That was medicine for Jay. She never questioned that it was the career she wanted, for all its rigor and layers of emotion. Yet she also never felt a full sense of belonging—especially not at the start of school.

On Jay's first day of anatomy lab in medical school, the students were divided into clusters of five or six, and each group got a cadaver to examine and dissect. Jay's instructor stopped by her group and introduced himself. Then he asked Jay and her classmates to share their names and where they went to college.

"Yale," came one response.

"Cornell," came another.

The instructor nodded vigorously both times. "Oh wow," he said, followed with a comment like "What a great school." Jay can't remember his exact words now, but she remembers the subtext: "You'll fit right in."

Then the circle came around to Jay. "SUNY Geneseo," she said, feeling her face flush pink. She was the only person in her anatomy group who'd gone to a state school and not an Ivy

League college. The instructor nodded and then, without comment, moved on to a nearby table of students. The lingering seconds of silence felt piercing. The others didn't seem to notice, reverting quickly to conversation about their next exam. They started dissecting their cadaver—a frail woman, facedown. This, Jay felt comfortable with; she had spent a summer before medical school working at a hospital, doing wound care. For weeks she had wrapped bandages around festering skin turned crusty with gangrene. At the start of the day's laboratory, she had been feeling confident in her capabilities as they prepared to turn their scalpels to the dead woman's flesh. But suddenly in this group of fancy degrees she felt useless, like an extra limb.

The exchange confirmed for Jay what she had known since she first decided to enter medicine: the field was designed to absorb certain people more easily. There were born-with-a-silver-stethoscope types. They were the kids for whom being a doctor ran in the family; they still had to do the work, but they didn't have to fight their way in.

For the rest of the dissection that day, she wondered if she seemed as misplaced as she felt to the rest of the group. She tried to focus on the task at hand, cutting their cadaver's back open from the nape of the neck downward. But she had only ever tended to the wounds of someone living, and turning her clinical tools to a dead person felt wrong somehow, dehumanizing. She told her mom this on the phone that night, and her mom reminded her that the woman had donated her body for medical research. Before going back to the lab the next morning, Jay wrote the word GIFT on her wrist, right above the skin that would be covered by a latex glove. She made it small enough so anyone would probably mistake it for a prompt to buy a birthday present, but noticeable enough that it would serve as a reminder: the body was a gift. The work had a purpose. Jay was meant to be there.

When Jay was seventeen, and excitedly told her family that she wanted to be pre-med in college and become a physician, she recalled her dad sitting her down and explaining that he wasn't going to help financially with her pursuit of a medical education. To his mind, this was for her own good. He told her that he wasn't ashamed of her career ambitions. But he was certain that she would one day wake up and reverse course, deciding that she wanted to settle down and start a family instead of putting in long hours at the hospital. Her dad had grown up with a certain set of expectations for women and men that he saw as indelibly linked to their happiness—the men would be breadwinners, the women would care for the family. So he knew, or at least hoped, that Jay would drop out of medical school, and any family money that he put toward her education would be a waste. Her two younger brothers, on the other hand, were a sound investment.

"This is just because I want you to be happy," he told her. "I'm not going to financially support things that you're going to regret doing."

Jay sat facing him in stunned silence. They had rarely spoken about her future in detail, but he seemed to take pride in her good grades. He enthusiastically attended her ballet recitals (even when she inexplicably insisted on a blue tutu), and cheered her on when she played soccer as the high school varsity captain. He seemed to support the industriousness she brought to her different endeavors. But those were childish pursuits. Now she felt ashamed—ashamed that she saw herself as qualified for the grit and strain of a medical career, embarrassed that she believed in herself when her dad's only plans were for her to marry and have kids.

"Thanks, Dad," she finally stammered, choking out the only words she could manage. "But you know," she mustered, "I'm going to do it anyway."

As Jay prepared to start college the next fall, on a 91 percent merit scholarship at the State University of New York at Geneseo, she spent her evenings and weekends working to save money for medical school expenses. She tutored biology and chemistry for

$60 an hour, and in the summer she lifeguarded at the local pool. In Jay's second year of school, her brother Christian started at SUNY Geneseo, also as a biology major, but the family fully covered his school costs without hesitation.

On the nights when Jay stayed up late, making flash cards, she felt a simmering anger. It wasn't directed at her dad so much as at her situation—the difficulty of the career path she had chosen, which seemed to demand endless time and money she didn't have. She felt a caustic sense of fear that she wouldn't make it to medical school; she also knew she couldn't succumb to her own doubts.

Jay's mom, Kate, was more supportive of her career path, though she worried about all the effort it demanded. When Jay called home senior year to say she'd been accepted to a medical school in Manhattan, her mom was so confounded that she just sat there in silence for a minute. In her mom's mind, med school acceptances were for people with money to hurl at MCAT tutors and application coaches.

But if you thought about Jay as a little girl, it made sense. Jay had always had a knack for jigsaw puzzles, persisting even when her siblings got restless. There was this rapture she seemed to exude when she studied for science class; for her fifth birthday she'd insisted on a *Magic School Bus*–themed party where her mom dressed up as the kooky science teacher Ms. Frizzle. And even as a child Jay had these caretaker instincts. When she was eleven, her youngest brother Jackson almost drowned in a swimming pool, and she managed to dive into the deep end and pull his head up as he sputtered and gasped for air. Scanning through these memories, like slides on a projector, Jay's mom could make sense of the medical career interest.

As the reality of medical school drew closer, Jay's dad began to waver on his initial opposition. A few weeks before her semester started, he called to say he could give her a check to put toward tuition. "Oh my God," Jay said breathlessly. "Thank you so much." She was excited about the money, but more so the phone call. She kept replaying her dad's voice when he made the offer. It was an investment. It was a concession that her career plans were more than murky delusions.

But on the Thursday of her first week, right before her first medical assessment, he called her to say he'd changed his mind. "I can't do it," he told her. He still felt uncomfortable with the idea of her pursuing a high-powered career. He said all he wanted was her happiness. The most clear, acute sense of happiness she'd ever felt was the day she got her medical school acceptance, but that seemed impossible for her dad to understand. She could hear in his voice how much he hated letting her down; he could barely formulate the words. She was thinking she would have no choice but to take out loans, as so many of her classmates would do too.

Three days later, Jay went for a stroll around Manhattan to clear her head. She was feeling like no one—her teachers, her dad—thought she belonged at school. Maybe she had made the wrong decision. She was standing on the subway platform at Forty-Second Street Grand Central, waiting for the 6 train, when she got a call from her aunt and uncle. They had heard about her dad's reversal, and they wanted to lend her the tuition money she needed.

Jay was so surprised that all she could say was "Thank you," over and over. She had to press her back to a subway pillar to stay upright, not even thinking about its layers of grime. She was overwhelmed. New Yorkers shoved past her in a hurry to get somewhere that must've been important, none of them paying any mind to the small blond girl gasping into her cell phone. Jay's eyes started leaking. Finally she felt like she could belong in this city of people moving and becoming, like her place could be earned and not just borrowed on interest.

At Jay's medical school, the annual graduation ceremony featured a certain tradition. Each student was instructed to choose one family member who came onstage and placed the customary hood over their head, signifying the transition to their next stage of training. Jay promised her mom, Kate, the role during her first year of school. Kate was so invested that she brought it up periodically, as if to make sure the privilege hadn't been revoked.

This was never a real concern; Kate was Jay's closest confidante, the first phone call when her daughter needed to vent, and the obvious choice to come onstage for the celebration. The two even looked alike—both blond with broad smiles, long faces, and angular jawlines.

But instead, Jay was notified in an email that she had officially graduated from her school, days before she was set to start her work in the Covid wards. There was no hood, no stage, just a digital formality. (The actual ceremony, to be held on Zoom, would come weeks later.)

Jay's feelings about the occasion were somewhat dulled by all the apprehension surrounding it. Jay had asthma, which meant she could be at greater risk than others if she got Covid-19. Some studies seemed to suggest it could increase the likelihood of hospitalization. Jay's mom was baffled that she wanted to join the group of early graduates heading into the hospital to work in the coronavirus wards. After all those years of late-night study sessions and stress over exams, Jay was finally months away from starting her career. It was the period of life that had seemed so implausible in high school, or even in her years at SUNY. It seemed to her family like maybe she should just relish these last weeks before work, instead of right away putting her life in jeopardy.

Jay's dad was uncharacteristically quiet about the decision. But her mom, who had taken such pride in her daughter's work ethic throughout medical school, phoned every relative and neighbor, hoping someone would know what to say to either put her mind at peace or help her talk Jay out of the idea.

Kate called up a friend who worked as a pediatrician in their town. (She was the first female doctor Jay ever met, and an important influence on her interest in medicine.)

"I'm absolutely freaking out," Kate said. "What do you think we should do? Should we not be letting her do this?"

Kate's friend said that it would be a learning opportunity for Jay, and that given the scale of this pandemic, Jay would likely have to care for Covid patients at some point, so it was better to start early. But there was a catch in the woman's voice. She, too, was worried, which only sharpened Kate's nerves. She wandered around the house that week in a zombie-like state. She tried to eat, but nothing went down. All the television anchors were forecasting death like it was the weather: there could be 2 million cases, 200,000 dead. Kate wondered if this was the way mothers felt when their kids were drafted. But no one was shipping her daughter off; Jay was enlisting.

Jay's brother Christian, who worked as a physical therapist in California, called his mom in a panic one afternoon. "I don't want her doing this," he said. "I'll give her all the money I save at my job."

"What do you mean?" Kate asked him.

"I just really don't think Jay should go in," Christian said. "I don't need much money, besides what I use for rent and food. If I give her everything else I earn for a year, will she not do this?"

Christian had always been the peacemaker in the family, placid and quick to appease, so his sense of alarm made Kate even more uneasy. (Jackson, the youngest, was tuned out of the family stress, mostly consumed by the disruption to his own college semester.) But none of them could talk Jay out of it. One night, she sent them an article about medical students from the University of Pennsylvania who graduated early to fight the Spanish flu, in 1918.

At any rate, Jay's decision to start work early had nothing to do with money. She felt an unwavering sense of pride when she thought about going into the Covid wards. Throughout medical school, students like her wandered the hospitals trying not to be a burden, wondering if they were qualified to help. Now none of that mattered. The city needed them. The sense of obligation was especially weighty for Jay, who had spent so long wondering if she even had the mettle for the profession. She kept glancing over proudly at the construction paper diploma her roommates had made for her, a star-shaped page that read "Congratulations!"

Kate wrote an email to a group of close friends on Jay's first day in the hospital:

As per an executive order signed by Governor Andrew Cuomo, Jay will graduate early today from medical school and accelerate her entry into the medical workforce due to the Covid-19 crisis.

It was supposed to be different. I was supposed to have the incredible honor of hooding her on the stage of Lincoln Center. I was not supposed to feel as if I'm sending her off to war.

She ended with a religious invocation: It is time to ask for prayers from family and friends. May the angels that helped clear her path thus far be forever at her side.

Jay tried not to think about these words too much as she prepared for her first day. For all her family's good intentions, she often made her best decisions by listening to her own internal voice.

Jay could remember with inordinate precision the first afternoon she spent in a hospital. It was the day she resolved she would go into medicine, to follow after those white-coat-wearing figures who saved her little brother's life.

It was on a skiing trip to Massachusetts when she was ten. Her brother Jackson, who was four at the time, was skiing twenty yards behind her. Jay turned at one point, and there he was, facedown in the snow, all limbs and winter jacket. She hesitated, but when he didn't move, she began to scream. Her dad turned around and yelled for her to get ski patrol. She propelled her body forward, downhill, grateful for how the command gave her a focus for her energy. *Get ski patrol.*

She waited as the patrol went off. They brought Jackson down in a sled. In the lodge, he screamed when they tried to remove his boots. Jay had to cover her ears. These were sounds she didn't think her scrawny younger brother was capable of making. Her mother's breaths came in huge, labored waves as her eyes bore into Jackson's face.

Her mom got in the ambulance with Jackson, and the rest of the family tailgated them all the way to the hospital. Jay's mom had a walkie-talkie, given to her by the medics, and periodically she relayed updates.

"His lips are blue," she said at one point. Her voice broke, as if willing the words to be untrue. "He's not responding."

The family's car went silent as they listened. Jay stared at the passing landscape, the sharp, endless white of the snow and the knobby trees reaching skyward.

Jay heard the EMT's frantic voice over the walkie-talkie. "Don't kill us," she heard him tell the driver. "But step on it." Then the sirens switched on, shaking those quiet Massachusetts roads. The ambulance leapt forward, and Jay's dad followed.

They took him to a small hospital, a few towns over from Baystate Medical Center, where Gabriela's Grammy worked. In the waiting room, Jay's dad paced in tight circles. One of the doctors came out and showed them an X-ray. "There's a fracture in his femur and severe bleeding in his leg," they were told. "He needs emergency surgery." Jay didn't know what that meant. But she could read the fear written on the doctor's face, and on her dad's. She wanted to ask for a translation of this clunky heap of words, but really all she needed to know was whether her little brother was going to die. (Looking back later, she would come to understand their looks of alarm. The femur is a strong bone and hard to break except in cases of severe trauma, like car accidents or serious falls.)

Hours passed. She sat on a firm chair, sweating through her layers of winter gear. Her dad paced. She closed her eyes and prayed that when she reopened them, someone would be coming back down the hallway with news. And finally, late into the night, the doctor reappeared, his pursed lips now turned into a smile: Jackson would survive. He would be able to walk. In Jay's mind the sight of medical personnel became linked indelibly with sweet relief.

Later, during medical school, whenever she was passing through a hospital's reception area, she'd think of those long, stretchy hours, silent and impossible to measure. She kept an eye out for girls like her, tiny torsos in uncomfortable chairs, waiting to know if they could get their family members back.

Six

ELANA, September 2016

Elana didn't know exactly what to expect of her first cadaver, but she didn't anticipate its manicure. The fingernails—trimmed, filed, and painted rose pink—were the second thing to catch her eyes; the first was the tumor coming out of the breast. The skin was slightly yellowed and rubbery, the result of formaldehyde pumped through the body's veins, but closer to a living human's than Elana expected. The hair was also well maintained, oily but free of knots, as if someone had recently run through it with a comb.

One side of the woman's chest was marked with a long, thin scar, where clearly she'd had a breast surgically removed, most likely in a previous bout of cancer. The scar and the tumor together seemed to form an obituary: one round of cancer she treated with a mastectomy, the next she succumbed to. She fought the disease, until she couldn't. But even at the end of her life, someone had been intent on polishing this woman's nails. The manicure was a reminder of how recently she'd been alive and cared for.

Whether the story she'd constructed for her cadaver was true, Elana couldn't say for sure. But it was necessary. Elucidating the cause of death and medical backstory, assuring yourself of family care—it all helped in making the first incision, reminding yourself that this dead body was once the living body of a person who wanted at the end of their life to be of scientific service. With the right narrative, the first cut turned from barbaric to benign.

Elana had once been told that you should treat every patient like family. That mattered to her, the notion that she was operating on a human with a story, even if that human was now dead.

What was most dazzling to Elana was the inside of the head, which they didn't dissect until later in the semester (the back, stomach, and chest all came before, so they didn't have to work on a decapitated cadaver). Inside was a maze of cavities and clusters—so many discrete parts that all functioned and synchronized. Elana felt a sense of wonder, piercing through the jadedness that had crept in during her early weeks of anatomy lab. She knew the inside of the human head was complicated, but she couldn't have imagined it had this many working pieces. She wanted to map all its parts like a cartographer.

Mid-semester, Elana got in an argument with a friend at Einstein about whether it was a waste for medical students to dissect these human bodies before they'd even finished a full anatomy course. These were precious resources, her friend argued, which could have been studied by more experienced medical researchers but instead were left for medical students to desecrate. This was jarring to Elana, raised in an observant Jewish family. The Torah explicitly prohibited desecrating human bodies, and she began to wonder if that's what she had done when she took to this woman's joints with a scalpel.

But as she examined the inside of the cadaver's head, that concern waned. What Elana was doing was studying the intricate construction of the body—the design of the skull, how the nasal cavity was built to humidify the air and trap its pathogens. That's the work of a higher power, she thought. Someone created that hollowed space. Someone sublime.

She paused her work for a second and thought of a prayer that she could say. "Blessed are You, Adonai, our God, King of the universe, who formed man with wisdom and created within him many openings and many hollow spaces," she whispered under her breath.

This was the prayer typically recited after you went to the bathroom, but it seemed to fit here. "It is obvious and known before Your Seat of Honor that if even one of them would be opened, or if even one of them would be sealed, it would be impossible to survive and to stand before You even for one hour," she continued. "Blessed are You, Adonai, who heals all flesh and acts wondrously."

Elana's memories of dissecting that corpse stuck with her, in all their curious, gruesome, and also beautiful detail. Mostly, she kept returning to this: Look for the pink manicure. Find the details that humanize a patient and remind you that they're more than a body on a table.

During the summer between Elana's third and fourth year at Einstein, she worked at an urgent care clinic in the Bronx. At the time she thought she would specialize in emergency medicine. She was quick on her feet and levelheaded in a crisis. She was technically skilled, adept at inserting needles and sewing up wounds, and more interested in understanding the full human machine than in any particular organ. She thought she would be well suited to the fast-paced nature of the ER.

Elana was working a late shift on July 3. The neighborhood was filled with the sound of early holiday fireworks, which lit up the sky outside like a Technicolor thunderstorm. The emergency room was packed, and the residents moved quickly between patients. Everyone was speaking at a rapid clip.

At some point in the evening, a middle-aged couple came in. They were directed to Elana, and explained that they had come because the man had a perennial nosebleed. It seemed benign.

But as the man described it, Elana noticed that his wife kept interrupting him. "Tell her about the last time this happened," she'd say, or "Tell her about that other problem."

Elana began to realize that this nosebleed wasn't just a nosebleed. The patient had a laundry list of other medical issues that he'd neglected because he didn't have a primary care doctor. He didn't speak fluent English, so he found it overwhelming trying to navigate health care systems. Elana started scribbling notes, interviewing him and his wife about his medical history. As they spoke, the line of others waiting for her attention continued to lengthen, but Elana didn't want the patient to leave without a plan for his next steps.

Later in the evening, her resident came to find her. "Why were you asking that couple all these questions? Were you going somewhere with that?"

Elana explained that based on the symptoms the man described, she thought he might actually have cancer. She worried that after he left the emergency room he wouldn't get follow-up care, since he didn't have another doctor. She didn't think it seemed right to clean up his nosebleed and send him away.

"Look," the resident told her. "I don't think that was a bad thing to do. But if that's the type of relationship you want with your patients, I'm not sure if emergency medicine is for you. That's more the work of internal medicine."

Maybe noticing Elana's expression of discomfort, he continued. "I don't want to criticize your work," he said. "But when people come into urgent care with a specific issue, you fix it and move on. That's emergency medicine. Otherwise the rest of the emergency department doesn't get seen."

In the moment, Elana just nodded and thanked her resident politely. She had already let herself imagine a whole career path in the hard-charged world of the ER. But she hadn't given as much thought to the relationship aspect of the field, or the lack thereof.

Emergency medicine was about quick diagnosis. It wasn't about disentangling the complex layers of someone's medical needs, or slowly building their trust in the health care system.

Elana liked the technical aspects of her job, but she also liked the psychological ones, the intricacies of relationships. That night, at the Bronx urgent care, she began to think more about the type of physician she wanted to become—one who cared for people in all their messy, complex parts. She didn't like the idea of her patients disappearing on her. She also didn't want to be a specialist, because she wasn't as interested in the idea of studying just one organ without understanding the rest of the human being (and human personality) attached to it.

Elana switched her specialty to internal medicine later that month.

Elana could think of a dozen reasons why she should graduate early and work in the Covid wards, and just as many why she shouldn't. On the one hand, Elana's draw to the field of medicine was about morality. Providing care when she was needed most seemed like an uncomplicated act of goodness. On the other hand, some Jewish laws were clear: you should help your family before you went looking for others to help. Elana's father was immunocompromised because of his diabetes, as was her grandmother, and she'd promised them both that she would run their errands so they could self-isolate.

Elana's grandmother used to be a nurse practitioner (she worked through the AIDS crisis), and she was convinced that rushing into the hospital was a bad idea. She was never restrained in doling out advice on medical matters; when Elana told her she wanted to become a doctor rather than a nurse, her grandma replied, "Oh, so you want to be stuck up?" Her grandma had a theory that doctors and nurses spoke different languages. Doctors tended to problem-solve by starting with science and working out toward the human in front of them, while nurses went the other way around.

"Never get involved with a crisis if you can avoid it," she told Elana, in her clear-eyed, no-nonsense tone. "You'll either get sick or sued. Neither one's a good option."

Elana's dad wasn't sure he agreed. Decades ago he'd driven tanks for the army, so he knew the steely resolve required for frontline care. "I'll be proud of you either way," he told Elana.

She called her mentor, Sharon, a pediatrician on Long Island who specialized in infectious diseases. But Sharon's best quality had always been her refusal to tip the scales when Elana had difficult decisions to make. "Make sure it's what you really want to do," she told Elana. "They'll work you hard, that's for sure."

Elana kept puzzling. She had a personal rule that she wouldn't make spur-of-the-moment decisions; when people rushed her to make a choice, it made her want to slow down all the more, to make sure she wasn't being shoved toward a poor judgment. And her husband, Akiva, thought the best idea was whatever Elana had last vocalized. "It seems like you want to do it," he professed enthusiastically one evening, as Elana was musing that maybe this was a historic moment and she should assume responsibility.

An hour later, when she worried aloud that maybe it wasn't necessary to put both of them in danger, he again concurred wholeheartedly. "More power to you, Elana, you know what's best," he said, his voice no less assured than it'd been earlier.

This wasn't Akiva's fault. He had an unabating admiration for Elana's authority on all matters. He'd been this way since they first got close, playing Dungeons & Dragons the summer after their senior year of college at Yeshiva University. Elana was assigned the role of dungeon master, and Akiva was a kung fu monk. He fell for her fast and hard. Elana was reserved; she figured that lots of the girls had crushes on bearded, olive-skinned

Akiva, who baked the doughiest challah. She didn't want to get her hopes up. Turns out she didn't have to worry; her romantic interest was reciprocated.

Elana wanted to know how others in her medical school class were weighing the Covid choice, so she texted Iris. The two had become friends singing in Einstein's a cappella group, the Lymph Notes, whose repertoire was mostly show tunes like "How Far I'll Go" from *Moana*, and "Seasons of Love" from *Rent*. Iris was decidedly the more musically talented of the two, Elana had realized during her friend's first-year rendition of "Winter Song" by Sara Bareilles and Ingrid Michaelson.

"What do you think I should do?" Elana asked her friend. Iris's situation was a little less thorny because she had Covid antibodies. Also, Elana had always sensed a strong innate set of values in Iris. While Elana liked to turn to books and rules and other people for guidance, Iris seemed to make decisions with a sense of sharp clarity and independence.

"What would fifty-year-old Elana look back and wish you had done?" Iris asked her.

That made Elana's worries suddenly seem thin and distant. "I think fifty-year-old Elana would say do it."

She joined the list of Einstein students graduating early. Later that week was Passover. It was her first time celebrating far from family; she was only comfortable seeing them outside when she dropped off groceries, not indoors for a meal. But the holiday's grandiose messages suddenly felt more real. Resilience, survival, sacrifice—none of these was theoretical anymore. Passover had all sorts of obvious parallels to their new reality. They were celebrating liberation in a time of real-life plague.

Elana was in the grocery store when she got the email notifying her of her hospital start date. She grabbed a pint of ice cream to celebrate. Akiva broke out in song: "Doctor, doctor, gimme the news," he crooned. "I got a bad case of loving you."

BEN, April 2020

Iris and Elana's classmate Ben had considered what it would be like to give his first serious diagnosis, the grim faces and muddled questions that might ensue. It came, finally, during his third year of medical school. He was studying a patient's MRI and saw an alien mass of gray on the brain's left side. It had to be a tumor. He quickly got hold of a neurologist, who confirmed the diagnosis, which explained the patient's memory loss, confusion, and fatigue. Ben was doing a rotation working alongside just an attending physician, with no resident, so the doctor told Ben that he should lead the conversation delivering the news.

What Ben hadn't anticipated when he presented this diagnosis was the stare. The patient's pupils widened and then, searching the group, landed on Ben. It felt as if his eyes were doing an archaeological dig, excavating Ben's gaze for any promise of reassurance.

To his surprise, the patient's wife was relieved. "It's a tumor?" She had expected dementia. Dementia was irreversible, she thought, but a tumor could be excised. Then she'd have her husband back.

Ben had to temper the woman's expectations. He explained that what she was hoping for wasn't realistic. Surgery would be high-risk. The woman's face darkened, but she didn't look away as she swallowed the news.

Months later, at the start of his fourth year, Ben gave his next tough diagnosis. The patient was in his early sixties. When he turned up in the emergency department, Ben's supervising attending instructed him to tell the patient that he had end-stage kidney disease and needed dialysis or a transplant. The patient broke down in tears. He had suffered years of sickness and even an amputation, he told Ben, but dialysis was the line he refused to cross. He didn't want to spend his life reliant on machines.

The patient jumped up and ran for the exit. Ben followed him to the doorway of the emergency department, and they stood there for fifteen minutes while Ben offered steady assurances.

Part of him wanted to look away—the patient's eyes were wild with something between panic and grief—but he forced himself not to. Finally the man's face softened, and he followed Ben back inside to wait for blood work.

These exchanges convinced Ben that medical care was partly in the eyes. No one wants jargon. Ordinary people aren't taught to interpret charts and CT scans, but they can read facial expressions. This wasn't the type of lesson assessed on med school exams—but it was fundamental.

This new knowledge began to weigh on Ben in the days before his virtual graduation. Isolated in Einstein's dorms on Eastchester Road in the Bronx, he puzzled over how he was meant to care for patients suffering from an alien disease without being able to see their faces, or show his own. Covid care required more personal protective equipment than all his specialty rotations combined: gloves, gown, face shield, surgical mask, and under that an N95. How was it fair to ask patients to trust him when they couldn't see him? He would be bent over their beds—taking medical histories, drawing blood—looking like a lanky cyborg.

One afternoon Ben read an article about coronavirus care providers who were taping photos of themselves to their gowns so that their patients could see what they looked like. He decided to print out a photo of himself and safety-pin it to his yellow gown. At least then his patients could look him in the face.

When Ben decides he is interested in a particular subject, he bears down on it with a ferocity bordering on obsessive. These days it's medicine, but when he was young it was railroad transportation. At the dinner table he used to rattle off facts about trains.

When they first moved to West Windsor, New Jersey, following their dad's pharmaceutical work, Ben insisted the whole family spend a day at their local station learning about the New Jersey public transit system. There's a photo of him and his younger sister

Jenny from that day. They're both sporting Yankees caps and brightly colored jackets, Jenny's in fuchsia and Ben's in bright red. Their facial expressions seem to telegraph their attitudes about the activity: Jenny is crossing her eyes and squeezing her hips in a goofy stance, while Ben stares stone-faced at the camera. This was how they were. When Ben took up a topic, it was only with unrelenting gravity. Jenny thought he might grow up to be a railroad conductor.

Jenny was cooking dinner in her apartment at Rutgers, where she was a senior studying exercise science, when Ben called one night.

Rutgers had just announced that classes would move online because of the coronavirus. The hours had since begun to bleed into one another. Jenny stayed up late talking with her roommates, then rolled out of bed in the morning in time for class on Zoom. She'd begun to wonder if the quarantine would last longer than a month and graduation would be called off. It was hard to imagine, though her world had quickly become something composed of impossibilities: boarded-up storefronts around New Brunswick, a fifty-thousand-person campus deserted.

"Hey, Ben," Jenny said, answering the phone.

"Hey, Jenny," he responded. "Uh, I had some news I wanted to tell you."

Einstein was letting his class graduate early, Ben announced. They would start working at Montefiore to treat Covid patients.

"Is it safe?" Jenny asked. "Are there enough masks?" She'd seen social media videos of hospital staff begging for more protective equipment and clicked through them with a mounting sense of concern for her brother, who she knew would jump at any opportunity to be of use in the crisis.

Ben promised her that the hospital had masks, at least enough for any direct interactions with Covid patients. "I had a question for you, actually," he continued. "Could I put you down as my health care proxy?"

"Your what?" Jenny asked.

"My health care proxy." Ben had decided, days earlier, that his little sister was best for the role, both because his parents were older and because Jenny was, like him, on the more rational side of the family. They'd already had plenty of conversations about Ben's views on medical ethics. He'd explained to Jenny why he wanted to be an organ donor, and insisted that she sign up too when she got her driver's license.

"Being my proxy means that if anything happens to me, the hospital would call you," he continued. Only Ben could orchestrate a worst-case scenario with the same inflection he used discussing options for takeout dinner.

Jenny's head was racing with uncertainties, but none specific enough to crystalize into a question. "Uh, sure," she said. "I can be your proxy."

"I trust you," Ben said. "I trust that you'd know what to do."

"Um, okay," Jenny replied, and added quickly: "You have to make sure you stay safe."

She realized that she wouldn't be seeing her older brother for several months. If he was going into the hospital he would probably have to isolate from their family.

Jenny thought about the icy evening she had spent visiting Ben in New York earlier that winter. On the subway ride over, she'd been trying to keep her mind off the reports she'd read of Asian people being beaten up or called racial slurs as the virus spread beyond Wuhan. But she'd made it to Ben's apartment safely, and the two of them had rented a movie (they both loved crappy rom-coms) and gone out for sushi. Jenny had enjoyed that evening, spending time in her brother's city dorm room, which was as messy as she knew to expect from Ben. It was a glimpse into his adult life.

Later that night, she pulled out her laptop and searched YouTube: "Coronavirus doctors." "Hospitals and coronavirus." She saw physicians, on the verge of tears, asking why they couldn't get their hands on more masks. She saw a video of a Covid patient being admitted to an emergency room, his lungs failing right there on camera. She watched doctors standing in hallways filled with stretchers and IV drips and tried to imagine her wiry, dark-eyed brother in their place.

Seven

SAM, April 2020

Sam and Jeremy agreed that after each of Sam's hospital shifts, he would do the Hot Zone Dance: enter their apartment and pivot to the Covid corner, the alcove off the entrance once reserved for soiled shoes. It was the only space in the apartment that their cat, Evita Carol, didn't deign to visit. Its latest additions were a box of Clorox wipes, gloves, and a brown paper bag full of masks.

There Sam would strip off his scrubs and deposit them with his backpack, badge, and bike helmet in a pile on the floor. He would then move naked through the apartment while Jeremy sat on the couch on Google Hangout meetings, trying not to laugh and otherwise dutifully ignoring him until he was fully scrubbed.

On Tuesday, his first real day of work, Sam went downstairs to get his bike. His shift would go from 7:00 a.m. to 7:00 p.m. He cycled up First Avenue toward the hospital on Bedpan Alley, the ten-block stretch that includes NYU Tisch (where Gabriela was starting work the next week), the Manhattan VA, and Bellevue.

Typically a hospital is not itself the protagonist; it serves merely as the setting for its doctors' heroic tales. Not Bellevue. One of the country's oldest public hospitals, it is perhaps its most storied.

Although most say Bellevue was founded in 1736, some trace its roots back even further, to an infirmary built by the Dutch in the 1660s to take in soldiers with "bad smells and filth." Under the British, that small infirmary became a modest two-story almshouse on the site of City Hall Park.

As the crowds of poor and sick in the city grew even larger, New York undertook the most ambitious construction project in its history to expand the almshouse, purchasing over six acres of land along the East River. In 1816, after a lag caused by the War of 1812, the new complex, known as Bellevue Establishment, opened with an orphanage, asylum, morgue, "pest house," prison, and larger almshouse, all guarded by a stone wall. (At some point, there'd been a plan to name the pest house New York City Hospital, but that provoked complaints from the trustees of the private New York Hospital, which served the city's elite.) Its construction costs had tallied up to more than $421,000.

The hospital was better known for its history than its architecture. The granite and brick structure sat on First Avenue and Twenty-Seventh. Its face was a large glassy prism that jutted forward—a five-story atrium built over an older, more elegant brick facade during an ambitious 2005 renovation. Walk inside the glass doors, and you could still see the original administrative building, complete with an 1880s gas lantern. In essence, three architectural endeavors all connected to form a single facility: the centuries-old hospital, the granite 1930s administrative building, and the flashy modern addendum from the early aughts. It housed nearly nine hundred beds.

The hospital took its unique character from both its national renown and its position as a last resort for New Yorkers turned away from every other institution. Historian Sandra Opdycke recalled learning about a Nobel laureate who returned to working at Bellevue after a hiatus during which he received his prize and was immediately greeted by a patient who asked him for a bedpan. "Bellevue's got the guy with the Nobel Prize, but he may well be asked to go get a bedpan," she said, laughing. "That's the tension that makes it so fascinating."

Over the decades Bellevue became a fixture of the city, its patient rolls dotted with notable names, like Eugene O'Neill, Sylvia Plath, Saul Bellow, O. Henry, bassist Charles Mingus, who wrote the song "Lock 'Em Up (Hellview of Bellevue)," and bluesman Lead Belly, who wrote the "Bellevue Hospital Blues." It sat at the inflection point of every health crisis that reached New York, wearing the scars of the city's disasters like an outermost layer of skin. When yellow fever, cholera, crack, and then AIDS came to the city, Bellevue saw them first. Now came a new scourge.

Outside the building, the trees were teasing the start of spring bloom; there were dogwoods, pear trees, magnolias. Some locals had tried to brighten the Bellevue area as Covid began its assault. They'd hung a paper sign that read "Thank You Health Care Workers" in bright marker on one of the black columns near the hospital's entrance, but the rain had washed out every other letter. The glass front doors had their own posters as well. One read "Heroes Work Here" in unfussy sans serif.

When disease ravaged New York in the 1800s, in an age before modern vaccines and the study of contagion, the stakes for front-line providers were even higher. In the 1840s, when Bellevue was overrun by typhus, so many resident physicians got sick that medical students were recruited to work as full doctors; after all, medical training standards at the time were lax. An 1852 annual report for the hospital celebrated the fact that it hadn't seen any of its own staff die that year. "It is a subject worthy of congratulation that we can [speak] without the melancholy necessity of paying an obituary tribute to . . . members of the House Staff who have fallen victim to the typhus fever."

Some medical providers, realizing the danger they faced, were known to flee when disaster struck. A whole genre of medical literature developed around the subject of doctors running from outbreaks, beginning with Daniel Defoe's *A Journal of the Plague Year*, published in 1722, which describes these fearful physicians as "deserters." It was with this grim notion in mind that the American Medical Association passed its Code of Medical Ethics in the mid-nineteenth century, mandating its members to treat the suffering "even at the jeopardy of their own lives."

Bellevue's doctors saw it all from their position at the city's flagship public hospital. In the 1830s, while the city's wealthy left town in droves, the hospital was overrun with immigrants and impoverished New Yorkers sickened by cholera. One New Yorker compared the waves of "well-filled stage coaches" fleeing the city in 1832 to those escaping Pompeii "when the red lava flowed." Cholera killed some 15 percent of the city's immigrant population but less than 1 percent of native-born New Yorkers. Then, as now, the most vulnerable were the epidemic's front line of attack.

The pattern held true a century later, in November 1980, when a thirty-four-year-old gay man walked into Bellevue feverish and short of breath. He was given a chest X-ray and lung biopsy. The results were stunning: he had pneumocystis pneumonia, a wildly obscure disease. His T-cell count, a critical factor in immunity, was plummeting. A few days later a drug addict with a heavy cough came in with the same condition. Both were dead not long after.

As the cases mounted, medical researchers began to call the disease Gay-Related Immune Deficiency, or GRID. The media titled it "gay cancer." The *New York Times* covered the spread in July: "Rare Cancer Seen in 41 Homosexuals." But Bellevue doctors soon realized that it wasn't limited to the gay community. The terrified physicians referred to the risk factors as the Hs: "homosexuals, heroin addicts, Haitians, and hemophiliacs."

"We thought of ourselves as the 'Fifth H,' " a Bellevue resident recalled in David Oshinsky's remarkable history *Bellevue*. "The House staff."

Bellevue saw more AIDS patients than any other hospital in the country. By 1985 a fifth of its patients were HIV-positive; New York then accounted for a third of the country's AIDS cases.

With so little known about how the virus spread, the American Medical Association put out a statement effectively allowing doctors to opt out: "Not everyone is emotionally able to care for patients with AIDS," it read. In other words, no doctor should be forced into service, despite medicine's "long tradition" of confronting epidemics. In New York, Gay Men's Health Crisis could barely cobble together a list of fifty private doctors who would see patients with AIDS, and a poll of 350 city dentists found that 100 percent refused to see people with the disease. Inseparable from the word *plague*, AIDS was often spoken of as partly disease and partly punishment—it couldn't be detached from the shame and moralizing attitudes directed toward the gay community. In 1986 conservative writer William F. Buckley suggested on the *Times* op-ed page that "everyone detected with AIDS should be tattooed in the upper forearm, to protect common-needle users, and on the buttocks, to prevent the victimization of other homosexuals." (This prompted parallels to the tattoos on Auschwitz survivors.)

But the AMA's statement made two exceptions: patients with AIDS would be seen in emergencies, and in the country's public hospitals, of which Bellevue was the most famous.

Bellevue staff members weren't immune to the fear of infection, and some confessed that they avoided visiting their patients with AIDS whenever possible. Along with their fear, they felt a sense of impotence. The medical school curriculum focused on life-saving interventions, but there was little they could do to save those sickened with AIDS. "I mean this is a place where people who had their legs chopped off in an industrial accident come," one intern wrote. "We're used to miraculously helping people."

There were no miracles in the AIDS wards. One day during

a grand rounds talk on a cancer commonly caused by AIDS, a senior physician stood up in frustration: "Thank you for this very nice lecture, Doctor," he said. "But why does NYU have to be the *Titanic*?"

The epicenter of the epicenter was Bellevue 17 West, an AIDS unit that was originally set up with ten beds but kept growing. It was segregated because of the special sanitation required for its patients' vomit and diarrhea (though the virus couldn't be transmitted in these fluids), and because they were especially susceptible to other infections in the hospital. Someone affixed a sign in block letters to a wall behind the nurses' station: THE ONLY DIFFERENCE BETWEEN THIS PLACE AND THE *TITANIC*—THEY HAD A BAND!

There were some medical school graduates who specifically chose Bellevue for residency during this period because they wanted experience treating patients with AIDS, but no one could know what it would mean for them to be trained in this environment. Dr. Saul Farber, NYU's chief of medicine, told the media it would be beneficial for the interns and residents, who would learn about all sorts of rare cancers and infections from treating people with AIDS. An NYU faculty member asked students to track the diseases they saw on rotation. Normally their logs might be filled with breast cancer or appendicitis. During the AIDS era, half described pneumocystis pneumonia. One third-year student rotating at Bellevue wrote that regardless of the symptoms he saw, diabetes or chest infection, he could only think of AIDS.

What the trainees learned, most of all, was how to care for the dying. They could treat one infection in a patient with AIDS, then see the same person months later for something worse. "Every day felt the same," one Bellevue resident wrote. "Legions of feverish, emaciated patients admitted from the emergency room." They were "buried alive" in plague, "a soul-numbing tedium of affliction and despair." Some compared the sick patients to images from Dachau or Biafra, struck by how young these fragile bodies were: "Witnessing your own generation dying off is not for the faint of heart." There wasn't much to offer in hope, but they tried to show their patients care through small gestures, like letting partners stay for an extra while after visiting hours.

Some doctors looked for silver linings. Working with the dying meant "you cut out the superficial, the bullshit," one resident wrote. "It deepens your humanity."

But the overwhelming sense was one of dread. "Imagine somebody completely miserable, covered with lesions, maybe scary-crazy," another intern wrote. "Now picture somebody right out of medical school, five feet tall and maybe a hundred pounds, poking at him with gadgets and needles, and you get a tiny sliver, minus the God-awful smells, of what it felt like to walk into the place."

By the time Sam and his classmates arrived at Bellevue, the somber uncertainty of the early pandemic weeks had begun to fade. The hospital wasn't waiting apprehensively for the Covid wave to hit; it had very much arrived. On their first day, Bellevue housed nearly four hundred patients with Covid.

The building was rearranged to accommodate the patient surge. In the ICU, IV stands were moved outside closed doors, so nurses didn't have to stand right next to infectious patients while treating them. Filtering devices were installed to turn regular hospital rooms into negative-pressure ones, which prevented the pathogens inside from escaping. Dermatology offices became workspaces for teams of residents. Stationed outside every room was a cart of PPE: gowns, shields, gloves.

Sam was assigned to a unit on the sixteenth floor, which would take both Covid and non-Covid patients. His team was composed of doctors redeployed from across the NYU system, among them a resident in psychiatry, a rheumatology fellow, and a gerontologist who typically staffed an outpatient clinic. Sam had worked at Bellevue during his medical school rotations, so he was well acquainted with its layout and workflows. He knew these even better than some of the nurses who had traveled from out of state to assist the New York hospitals. But because he had been away from the hospital for several weeks, he also had to remind himself of the basics, like how to navigate the electronic health record system, order blood, print out labs, and make sure those labs got processed. It was like relearning how to ride a bike, feeling wobbly at first but then steadier the less he worried about it.

Sam's first day was charged with the electricity of initiation. He was enacting the series of real-doctor motions that for years had sat in a distant, future realm. There was the first time someone on the phone called him Doctor; the first time he put in a medical order without needing approval from a superior. (One of his early orders was just for Tylenol. A more practiced doctor wouldn't even have blinked. But Sam heard the words like explosions in his brain: *I'm giving Tylenol! For pain!* The training wheels were off.)

Some of the early graduates' worries about protective equipment shortages were assuaged when they started work. Contrary to reports they had heard that doctors could visit their patients wearing just surgical masks and eye protection, they were informed that nobody could enter a patient's room without an N95. Hospital staff understood that they were all vulnerable to one another; one person gets exposed and becomes a force multiplier.

Sam was told he would be given around seven patients to manage. He would have to weigh prudently when to visit them in their rooms, because each trip to the bedside required an N95 mask, face shield, and gown. Precious resources. The early graduates were told to minimize their in-person time with patients whenever possible.

Sam had a few patients he visited in person those first few days, stopping by if he needed to draw blood or check their urine output. He usually waited until later in the day, so he could take care of all the tasks he needed to do for them at once. He took the opportunity then to FaceTime their families, too. As he balanced his phone in front of their faces, he was acutely aware that he was exposing himself to the virus. Every minute in the room made him feel more radioactive.

Two of his first patients were middle-aged women, both immigrants from South American countries now living in Brooklyn and Queens. One of the women was desperate to be discharged. She had been given a breathing tube, and when she finally started regaining some capacity to speak, the first thing she communicated was that she had to leave the hospital; she was worried about the mounting financial cost. In her mind, every week at Bellevue would be another rent check. Sam knew that her bill would be capped; the cost could still be catastrophic, as she had feared, but at this point going home early likely wouldn't help.

Sam had other patients that week whom he couldn't meet in person at all. There was no clinical reason for him to enter their rooms, so he orchestrated their care from afar. This meant a peculiar rewiring of hospital instinct. Instead of stopping by their beds whenever he could to ask about comfort levels or even just to say hello, Sam studied his patients' charts. He gathered information and passed it along to his attending physician or senior resident, who would briefly visit patient rooms whenever they had to make some change in procedure or medication.

Soon he knew everything about these patients on his list: their medical histories, allergies, vital signs, pain medication dosage, Covid test results, dates of birth. He just couldn't see their faces or introduce himself. It was difficult to get a sense of the human body and personality attached to the labs he was reading. Meanwhile, the patient, alone in bed, might not even know Sam existed.

One such patient, taking shape in the charts on Sam's screen, was young, and dying. He had a spinal tumor, a form of cancer as aggressive as it was rare. He hadn't tested positive for Covid, but every aspect of his care was structured by the virus. Doctors were minimizing time in his room. His family couldn't visit. His tumor had metastasized quickly. With its rapid spread, he had little chance of long-term survival.

Sam's senior resident set a time for the team to discuss the patient's goals of care with his oncologist. In the absence of conversation with the dying patient, Sam wanted to know everything about him—it was his patient, after all. He read about his response to various new therapies, wondered how he was coping.

Normally, meeting a patient's visitors gave you context. You walked into their room and found the parents writing out lists of clothing items they should bring from home or reading out plays from the latest Jets game, the confused younger sibling dumping juice on the hospital floor. The patient transmuted from a series of medical procedures into somebody's doted-on son or favorite older brother. All that evaporated in the coronavirus age: no visitors.

Sam wondered if there was any appropriate way to stop in the patient's room for ten minutes of conversation—though the patient had no idea who Sam was, didn't realize there was a young intern down the hallway reading his charts. Maybe Sam just wanted the certainty of detail. What did it mean to fully care for someone without speaking to them? Sam wondered if he could even walk by the patient's room and wave through the door, but he wasn't sure that felt appropriate.

Finally, when the attending physician was out over the weekend, Sam decided to visit the patient's room. In the hallway he slipped on his gown, N95, and shield. He stepped inside and realized instantly, as he made eye contact, the chasm between the piercingly detailed image of the patient he had developed and the actual reality of a human being in front of him.

Sam introduced himself, explaining he was the intern on the team but hadn't been able to come by because of the efforts to conserve PPE.

The patient stared at him blankly. "Do you have the results from my MRI?"

Sam actually did have the MRI results. But he knew that the patient's pathologists, neurologists, and oncologists were all meeting the next week to discuss his care plan. Since the experts would have more definitive information for the man within a few days, he didn't want to potentially misinform him. "We have the scan, but to know what it means for you going forward we have to wait for the tumor board," he said. "Which isn't today."

Sam was quickly settling into the strange new realities of coronavirus care. He knew that there was no way to shortcut a patient relationship from behind a workstation computer.

In the electronic health record system, this patient was a series of labs and drugs. But off-screen, he was a young man in a solitary hospital bed, wrestling with the prognosis delivered by masked faces who said he might not have much longer to live.

Sometimes, from their charts, the Covid patients appeared to be mending. Then suddenly they crashed. It was often day seven or twenty-two of their illness; those were hot spots.

It was Sam's first week on night shift, and he heard the call over the PA system. "Airway code, floor sixteen." A code meant a patient had become unresponsive, either through cardiac or respiratory arrest. An airway code activated the anesthesia team as well, meaning the patient might have to be intubated. If they heard the words "rapid response," that meant a patient's conditions had abruptly deteriorated and they needed immediate evaluation by physicians or nurses, who would decide whether to move them to the ICU.

Again, Sam wouldn't actually enter the coding patient's room, though he could help by rushing to the scene of the emergency with supplies. Only a handful of residents would go in and do chest compressions, which meant pressing on the patient's chest to distribute blood to critical organs. Sam flipped down his face shield and sprinted down the hallway. He grabbed a LUCAS device, a large ring-shaped silver apparatus used for mechanical chest compressions, and extra PPE just in case.

Sam's role at the code was, as his friends put it, the computer guy—he had to stand right outside the patient's door with a computer and get pertinent information out as quickly as possible. Did the patient have allergies? What medications were they on, and when were they last given? Were their family phone numbers in there? What was their code status? That was told in three telltale letters: DNR meant "do not resuscitate," and DNI meant "do not intubate." Full code meant they wanted both resuscitation and intubation, any possible intervention to save their life. In normal times, Sam might have brought the computer inside the patient's room, but now that would've entailed unnecessary risk.

As residents gowned up to go inside, Sam passed the information along to them. His eyes locked on the computer screen as he listened to the doctor's voices ebb and flow on the other side of the wall: "Continue compressions," and then "pulse check."

Then he heard: "Pulseless."

"Time of death?"

From the other side of the door, Sam's eyes moved back over the biography on the screen in front of him. All he really knew about this person was her code status and the last time she was given her meds.

The residents leading the code held a short debrief: what went well, what hadn't gone well, what they could do differently. For this code, the outcome had come so quickly that there wasn't much to discuss. And Sam had borne witness without even physically seeing the woman who died.

Other medical students, growing up, said they watched television shows that made doctoring seem like an activity that comes in short bursts. A patient is flatlining. A doctor careens onto the scene to begin compressions, quick and reflexive. A patient on an operating table is still, unconscious while the surgeon expertly maneuvers the scalpel toward open flesh. In these moments, the doctor is heroic, the patient cardboard, dull and lifeless.

In the hospital, these mock-ups of medicine could begin to seem ridiculous. Medical trainees saw how much of the work unfolded in long interactions complicated by patient personalities, by the fatigue of a late-night shift, by the nerves formed in an emergency scenario. Every procedure was less about the singular hero than the team, the people supplementing one another's skills and leaning on one another for support.

But if your job was to stand behind a computer during a late-night code, you might feel like an extra on a medical TV show. The action unfurled behind a wall—residents calling out commands, costumed in shields and gowns, a lifeless body beneath their hands.

Eight

IRIS, March 2020

As Iris started reading news about the virus, she found herself thankful that she'd been randomly assigned to pulmonology in Paris. She'd already spent weeks studying the lungs.

Forty-eight hours after she landed in New York, she developed a dry cough. She didn't think anything of it. She felt a little fatigued, but that was likely a combination of jet lag and the tumult of moving her life and belongings back across the Atlantic to the Bronx. It wasn't clear yet what Iris and her fellow fourth-years were supposed to do now that they weren't permitted in the hospital, so she was trying to approach her days like a pseudo-vacation. On Tuesday she and Benjamin ordered tacos for dinner.

"Man," Iris said as she took her first few bites. "These tacos are incredibly bland."

"I think they taste good," Benjamin replied.

"Seriously," Iris continued, chewing on her second taco. "These taste like nothing to me." She took another slow bite of battered fish. Beyond its crispy texture it could've been anything. Chicken. French fry. Mozzarella stick.

"Wait." Benjamin paused. "I was reading today that one of the symptoms for the coronavirus is losing your sense of taste."

"I haven't seen that," Iris said.

"It's in the French news," Benjamin said. He pulled up an article in French on his phone and passed it over to Iris. "Look—loss of taste and smell."

"What do I do?"

"Do you still have your sense of smell?" Benjamin asked in his characteristically nonchalant tone.

Iris knocked back her chair and ran to the kitchen. She grabbed a baggie of garlic off the counter and buried her nose in it. "I can't smell a thing."

She began to replay the last few days in her head. There was the cough that she'd ignored, and the tickle in the back of her throat. There was the exhaustion that she'd chalked up to travel. And then there was the fact that after weeks in an international hospital, Iris had gotten on a plane as packed as a Berlin nightclub with germy tourists. There was no doubt she might've been exposed.

What was disquieting to Iris was the novelty of the situation. After years of medical school, she wasn't used to getting sick with the unknown; she was used to having answers. She wondered whether she would wake up even sicker the next morning.

The following day Iris and Benjamin drove out to New York's first drive-through testing site, which was in New Rochelle, at that point still conceived of as the epicenter. Iris felt like they were on the set of a horror film, something about an alien invasion. They drove past a row of people in hazmat suits and face shields, and Benjamin shouted their names and address through the car window. Then they lowered the window for just thirty seconds while one of the extraterrestrial-looking figures plunged a swab deep into the recesses of their naval cavities, so far up it felt like the cotton was brushing Iris's brain.

It was one of the first moments when Iris felt acutely the gap between the way these weeks were supposed to shape up and their reality. She was supposed to be in Paris. Instead she was in Westchester, with a stick swirling up her nose.

The two of them got their results a few days later: they were both positive. Iris was beginning to feel relieved. She still had nothing but a cough and loss of taste and smell. She realized that maybe the two of them were lucky cases, the ones she'd read about where young people were virtually asymptomatic but wound up immune. It would make her even better positioned for hospital work, if the school changed its mind and let her go in.

Iris decided not to tell her parents right away. They were worriers. And anyway, she figured there were only two options: she would recover quickly and then pass on the good news of her immunity, or she'd wind up in the hospital, in which case they would find out her diagnosis regardless. Besides, Iris's family lived down the street from Elmhurst Hospital, in Queens, so their days were already filled with the unease sparked by unceasing sirens.

Toward the end of the week, Einstein's administration emailed Iris and her classmates with news. They had the option to graduate early and begin work at Montefiore, possibly in the Covid wards. The details were still being worked out. Iris was elated; she could give the news to her parents in an inextricable, almost inscrutable blend: she was going into the hospital for work, but she was most likely immune to Covid. (Or at least *thought* she might be immune. There wasn't enough research for her to know with certainty.)

She presented the notion of early graduation as though it was mandatory: "They're sending us in," she told her mom. Like it was a directive, no questions asked. "Also, I caught coronavirus, but it's been mild."

Her mom was so shocked that she couldn't pick which piece of news to react to first. "You have *coronavirus*?" she asked, saying the word like it was an expletive. "And you didn't say anything?"

"I'm on the tail end of it!" Iris assured her. "Anyway, I get to help out in the hospital. They need more doctors right now. I think it's good that I can be useful."

For two weeks, Iris isolated in her four-hundred square-foot apartment with Benjamin. In the evenings they cooked and giggled over YouTube videos. During the day, Iris completed video training in preparation for her start date at Montefiore, which would come later in April.

On April 20, her first day of work, Iris's alarm rang at 6:45 a.m. She rolled out of bed quietly, trying not to wake Benjamin, and stumbled to the bathroom to brush her teeth. Normally she would put on makeup, but for day one on the Covid wards she figured it wasn't necessary. Her face would be behind a mask and shield—the hidden benefit of infectious warfare.

Forgoing the cosmetics gave Iris enough time to walk to the shuttle instead of her usual sprint. It pulled off campus sharply at 7:30, drove down Eastchester, and arrived at the door to Montefiore Medical Center's Moses Campus at 7:45.

Iris waved toward her friends filing in. Some of them had fled the city to quarantine with family in the amorphous weeks since the pandemic came, and it felt nice to be reunited; they'd seen each other only on frenzied FaceTimes and Zoom calls as they tried to figure out what to do about graduation and hospital work.

Upstairs there was a breakfast spread of fruit, yogurt, and eggs, which they could eat by briefly removing their masks. The residency program director came by to welcome the group. She read off everybody's names and gave a little cheer when she got to Iris's, as she was the only one of the early graduates who would be staying on as an internal medicine resident at Montefiore after the group's temporary frontline work. "We like you all equally," the program director said. "But especially Iris! Welcome!"

Next they had a tutorial on donning and doffing their personal protective equipment. The group gingerly accepted the gowns and masks handed out; their social media feeds had been filled with videos of medical workers pleading for their own PPE. There was the ICU nurse in Illinois who had worn a single-use mask for five days, and the ER doctor in California who had stored dirty masks in plastic containers. There was the video that showed a series of doctors in scrubs begging for help: "I'm on the third day of wearing my N95," one said, choked up. There were rumors of physicians and nurses at other New York hospitals using trash bags instead of gowns. The hospital administrators kept assuring them they'd get the supplies they needed. But it was hard to square with what the new doctors were reading in the news.

As Iris followed the sequence of steps—N95 on, surgical mask over that, then goggles and gloves—she thought about how any small slip could leave her exposed. A mask just a bit too large, a gown left slightly askew. At this point she might be immune, but it was too early to know for sure. The doffing was even more complicated. You had to delicately lift the mask up and get all the layers of protection off without touching the infected surfaces and virusing everything up. Iris also wondered how her face would handle the constant itch of the mask.

After the training, the program director sent them off to meet their teams. The patients on Iris's floor would fall into two categories: non-Covid patients with standard medical issues, like heart failure exacerbation, and stable Covid patients. Eventually she would handle eight cases at a time, but she would start off slowly, with just three or four. She would be assigned the less severe patients, at least for the first week, since every medical order she entered into the computer system had to be reviewed and approved by a resident.

One of her first patients had been transferred to the unit that morning. A month ago he'd been admitted to the hospital and swabbed for coronavirus. The test came back positive. Since then, his lungs looked to have been making a recovery, though he also had kidney problems and needed dialysis. She began to read through his chart, sitting at the nurses' station in the middle of the floor, preparing to introduce herself. Unlike at Bellevue, the Montefiore interns would be expected to visit their patients at the bedside right away.

Shortly after, a voice on the PA system called for a rapid response. Iris froze; the call was for her patient. He was marked full code—meaning that if the situation called for it, he would be intubated.

Residents started rushing toward the patient's room. Iris had been emotionally steeling herself for her first day, but she hadn't expected that her first meeting with her first patient would come as his lungs were failing. She wondered if she was even supposed to show up at his bedside. He was possibly about to be intubated and moved to critical care, which meant he would remain Iris's patient for all of the next few hours. Should she enter his room? That might be the most responsible choice or the least rational one—she could make a case either way.

She froze next to the nurses' station, her breath heavy behind the mask. Here was her warm welcome to the front line.

The road to a doctor's first day is an arduous one. There are the four years of medical school, with classes spanning physiology, biochemistry, molecular biology, and pathology. That wide range makes for a formidable course load. Most medical schools score the early years of classwork pass/fail, and students have a saying they repeat like liturgy: p = MD, a pass gets you a degree. But just scraping by isn't entirely an option, because at the end of all those courses comes an over-eight-hour licensing exam, whose result dictates where students can do their residencies. (The last-minute effort to get into a residency on the part of those students who don't "match" into their preferred programs used to be known notoriously as the "scramble.")

And after the stress of classes and exams comes a new form of pressure: the clerkships that bring students into hospitals to practice their clinical skills. They are scattered to different facilities around their city. After spending the first two years of school synced up—eating, sleeping, studying all together like boarding school—they are all on their own. Some rotations are known to be more grueling than others. Obstetrics and gynecology means twelve hours on your feet, starting at five or six in the morning. There are lots of moms in agonizing pain and not much hands-on involvement from the students, except in delivering the placenta. But at least clerkships offer the chance to learn the hospital's rules and rites before the first phase as a real doctor, residency.

The origin of today's American residency system lies in a reform movement much like the one that changed med schools themselves. It started at Johns Hopkins in 1889, when the hospital decided to assign recent medical school graduates to a number of clinical departments, including medicine, surgery, and gynecology. The hope was that these recent graduates could function as apprentices of sorts. At the time, there was an obvious need for medical school graduates to get more hands-on, supervised experience. Many were so poorly trained that it was common for surgeons to mix up a pregnant woman's fetus with an abdominal tumor and try to excise it. ("Doctor, there won't be any operation today," reported one nurse in a Pennsylvania hospital, in 1882. "The tumor was born last night.")

Proud graduates of the Hopkins residency began to export the system nationwide. Hospitals soon found value in the residents, too. With medical knowledge advancing in the late nineteenth and early twentieth century, average hospital stays dropped (from eighty-one days in 1855 to nineteen in 1899 at the Massachusetts General Hospital), but there was more to do in the management of patients and their pathology, from urine tests to bloodwork. Residents filled a gap.

Through the 1920s, there remained little standardization in the residency systems. Every hospital program had a different structure, duration, and level of responsibility it assigned to its residents. A 1940 report on graduate medical education said of these programs that "no two are exactly alike." Like the variance in medical school quality a generation earlier, this bred problems for patients. Top surgeons maintained that their craft required a minimum of a thousand hours of training—yet some people entering the field had no more than one week of practice, and that spent operating on dogs instead of humans. In the late 1930s, the Advisory Board for Medical Specialties determined that as of 1942, any candidate for certification would have to have completed a residency of at least three years at a satisfactory institution. With that, the wild, wild west of medical training finally began its transformation into the rigorously structured system students undergo today.

The first month of the first year of residency, called intern year, begins in July. In Britain it's called Killing Season. It is widely considered the most dangerous time to be a patient because the hospital has just initiated its newest staff. The fresh crop of doctors is overworked, loath to burden their superiors with questions, and there's always the chance of patients bearing the brunt of their inexperience: mixed-up vitamins or, more serious, an extra zero on the end of a dosage order. It was only in 2003 that residents' workweeks were capped at eighty hours nationwide, where before they sometimes went over one hundred. (Capping resident hours wasn't a clear-cut decision in terms of patient health; working too many hours was risky, but so was the period of hand-off, when patients passed from one shift to the other and critical information might be lost.) Whether the "July effect" holds up empirically is more controversial. Some call it an urban legend, but others say that teaching hospitals do seem to see a slight lengthening of patient stays in that high-stakes period of summer.

It's a rough time for the new doctors too. They have already invested so much time, sunk in so much money. For years, as students, they watched their attending physicians carefully and took mental notes on the types of doctors they might want to become. Then comes the day they are no longer watching but becoming. Their choices suddenly hold the weight of medical orders. It's a power that's hard-won, gratifying, and terrifying all the same.

Iris put on a gown and N95, but before she could even enter her first patient's room, the senior resident stopped her. There were already ten people inside, ensuring that the patient got sedated, intubated, and sent off to the ICU. Iris returned to the nurses' station. She spent her next two days tending to the less severe patients on the floor, many of them ready for discharge.

Her next seriously ill Covid patient was admitted that Thursday, around 4:00 p.m. Scanning his medical chart, Iris saw that Mr. Lopez was in his early sixties and of average weight. His oxygen saturation levels were low, and when she stopped by his bed, she could see that he was hyperventilating.

Mr. Lopez didn't speak any English, so Iris called a Spanish interpreter on the phone and introduced herself. "I'm Iris," she said brightly. "I'll be taking care of you." She was struck right away by the warmth that Mr. Lopez exuded, even with his face mostly concealed behind a mask.

From her weeks in the hospital in Paris, where she herself was often the one at a loss, Iris was especially attuned to patients who couldn't understand English. She knew how paralyzing it was to hear a flood of indistinguishable words, all spoken seriously, their meaning connoting either recovery or disaster. It was like sitting in front of a jigsaw puzzle where the segments were all the same color. The larger whole was there, but impossible to riddle out.

Iris watched as Mr. Lopez fought through quick, shallow breaths. She took his medical history and discussed the onset of his Covid symptoms. Then she had to get his code status.

"If your heart stops, do you want us to do chest compressions?" Iris asked.

Normally this would be a lengthy discussion, but they were communicating with a phone interpreter, through their PPE, and he was struggling for breath. It was difficult for the interpreter to make out his words. Iris had to focus on getting the essential information.

"Yes," Mr. Lopez told her. "Please do everything."

It seemed like at some point he might need to be intubated. But her team had determined that it was not necessary to do such an invasive procedure right away, one that would no doubt make his recovery longer and more complicated—if he recovered at all. There were serious risks associated with intubation. Instead they decided to see how he fared on a nasal cannula, a tube going into his nostril, coupled with a non-rebreather mask, which fit over his mouth and nose and connected to a reservoir bag filled with a high concentration of oxygen.

When Iris left that evening, around 8:00 p.m., Mr. Lopez was statting, meaning he was oxygenating well enough with the nasal cannula. Iris decided that she would FaceTime his family the next morning so they could say hi and see how he was doing.

That night around 10:00 p.m., in bed, Iris checked his chart on her phone and saw that his condition was deteriorating. His oxygen saturation level was continuing to dip. She wondered whether he would have to move to the critical care unit in the morning, and she considered how she might communicate the news.

Iris had studied how to deliver bad news in her third year, paired up with Elana for the simulation. The scenario they were given was to tell a patient who had just finished imaging that he had cancer. They each practiced on a medical actor, observing one another and giving critiques.

There was a script to it, but it was more of an improvisatory art. They were taught to rely on questions to make sure the patient stayed with them for each progressive step. Iris liked practicing with Elana because her friend seemed to understand the less mechanical aspects of it.

Elana—who wore colorful headwraps over her wispy brown hair and had doe-like, expressive eyes—meted out her words with the steady rhythm of a heart monitor. She had this careful choreography in the way she talked, as if she could predict exactly what questions would follow her every sentence. She knew not to assume that the patient had already accepted the severity of the situation; people tended to interpret from their doctor's words what they wanted to hear. Elana's role as the breaker of bad news was to prepare the patient for the blow with each successive sentence, like the "Can we talk?" text before a breakup.

"So you know we just ran this scan," Elana said, facing the medical actor. "Can you tell me about what's going on with you that we needed to do this?"

The medical actor was good. Clearly he had sifted through his file of personal and medical information, lined with details as granular as a historical biography. (The actors weren't all in this category; some forgot their fictional names or described their symptoms of psychosis with all the apathy of a car insurance robocaller.)

The actor's pitch teetered between mourning and denial. Elana's voice never shook. Iris saw her friend's eyes warm with something deeper and more textured than pity. Elana later remembered that the actor turned to her at the end and told her he'd never seen anyone navigate the conversation so seamlessly. Apparently she was highly skilled in the bad news department.

Iris was especially taken by her friend's directness, how she could speak the unthinkable—you're dying—without stammering. This wasn't the nature of communication in Iris's family. When Iris's grandma was dying of cancer, her mom refused to say the prognosis aloud. Even when the family went for a last visit, her mom insisted they were just stopping by to say hello. The memory of this time had left Iris with a hollow feeling in the pit of her stomach whenever she had to tell a patient something they didn't want to hear. Their textbook had a module on these types of conversations, but Elana made it all seem natural and doable. Maybe even replicable.

That Thursday, thinking about Mr. Lopez, Iris's mind returned to the simulation. If she had to give bad news the next day, she could channel her inner Elana.

Iris arrived at 7:00 on Friday morning, anxious to check in with the night team about how Mr. Lopez had fared overnight. As they gathered, the night nurse gave Iris a look she recognized. It was the same expression that had been on Elana's face during the cancer simulation.

"We have some unfortunate news," the nurse said. She turned to Iris. "Mr. Lopez passed away last night."

Iris froze. She prayed, desperately, that she wouldn't start crying—not on her very first week of work, her fourth day on the Covid wards, surrounded by physicians and nurses who had been on the front lines for weeks. Looking around, she saw that her team's faces, too, were all drawn in quiet grief. No matter how many deaths piled up in these wards, the new ones never got easy, whether you spoke with the patient for weeks or, like this one, three hours. And for Mr. Lopez, the sense was that there wasn't anything more they could have done. Knowing what they knew yesterday, and understanding the low survival rates of those on ventilators, intubating him hadn't seemed like the right choice.

"Let's take thirty seconds of silence," the attending physician said.

Standing at the nurses' station, Iris suddenly wondered whether she had made a mistake coming here. These other doctors and nurses had no choice about serving Covid patients, but Iris had actively enlisted. She'd decided she was capable of taking on all the anguish that entailed. She felt overpowered by a blistering sense of loss—but shouldn't she have been more emotionally prepared? Maybe she had no excuse to feel so paralyzed. She'd come here entirely of her own volition. She wondered, too, whether there was something else she could have done for Mr. Lopez in his last day. Why the hell hadn't she put his family on FaceTime yesterday?

This was Iris's first time losing her own patient. She bit her lip, steadied her breath. Then the moment of silence broke. The night team continued with their handover notes. And Iris steeled herself for her next twelve hours of work, and for all the days and months after.

ELANA, April 2020

Elana consumed books insatiably. Sometimes she finished two in a week. She liked Agatha Christie mysteries, crime novels by Dorothy L. Sayers, stories of heists. Revisiting Jane Austen never got old. Elana did her best reading on Saturdays, when she would take her books to the quad outside the medical dorms if the weather was nice enough. That was the one day she couldn't work, in observance of Shabbat. Elana's family members all carved out twenty-five hours without any of the normalities of the weekday: technology, money, telephone, electricity, driving, shopping, laundry.

Elana also liked to spend her Saturdays nature-walking. Her mother, raised in northwest Michigan, had taught her to forage and identify plants in the wild, and Elana could spend hours strolling by Einstein and looking for mulberries and huckleberries. Most people didn't realize that the Bronx was full of mulberry trees. But Elana's favorite were the huckleberries, because when you turned over their leaves, they reflected the sunlight like tiny disco balls. The streets were overrun with garlic mustard too, a weed brought over by European settlers that tasted like seasoning if you ground it up.

Elana had always appreciated the rigidity of the Shabbat rules, especially the excuse they offered for long meals with family. She didn't mind them even in medical school, when her classmates spent the weekends studying or going to bars. She liked the structure they gave to her weeks, the demarcation between one and the next. The gift of Shabbat was the reminder to separate your being from your doing. The Jewish scholar Abraham Joshua Heschel called it "a cathedral in time": "There is a realm of time where the goal is not to have but to be."

But the Shabbat rules had seemed less complicated when the work that Elana put on pause wasn't lifesaving. As her start date at Montefiore approached, she called her dad to ask whether he thought she should observe Shabbat even when the hospital needed her.

"Now that you're a doctor, your religious obligation is to save lives," he told her. He reminded her of the Jewish principle *pikuach nefesh.* This meant that saving a life trumped any other religious commandment, including the rules of Shabbat.

If Elana was going into the hospital during a pandemic, and taking on all the risk that entailed, she wanted to be maximally useful. She decided that if the hospital called her in on a Friday night or Saturday morning, she would work. She explained her reasoning to Akiva in words that she knew were equally aimed at assuring herself: "At home, Judaism is my religion. At the hospital, it's medicine."

But on Elana's first Friday evening at Montefiore, realizing she might not make it home before sundown, she felt a wave of dizziness. It would be her first time using a computer on a Friday night, her first time getting in a car to drive home.

Shortly before sunset, Elana stepped out of the workroom and pulled out her phone. "This is making me feel weird," she texted her dad.

He texted back immediately. "You gotta do what you gotta do," he wrote back. "When I was in the army, they had me driving tanks on Shabbat." She could almost see the subtext of her dad's words, like notes written in the margins of a book during Bible study: *pikuach nefesh*. They were lifesavers, the two of them, swallowing their fears.

As the sky outside the hospital darkened, Elana had to keep repeating to herself: *Saving a life trumps the Sabbath.*

Nine

JAY, April 2020

"Good morning." The words were muffled, coming from the masked mouth of the woman behind the hospital's front desk. "How can I help you?"

"I'm here for the new doctor orientation," Jay replied.

"Take some hand sanitizer," the receptionist said. "And please put this on." Her gloved hand passed over a surgical mask.

Jay was already wearing a mask and had sanitized moments earlier, but not wanting to be a nuisance, she dutifully complied, placing the new mask over her first. Then she took the elevator upstairs, where a group of six other early graduates sat waiting for orientation to begin. Following the social distancing bible, they were scattered around the room at a distance of at least eight or ten feet from one another. Soon they would start to visit patients, and the luxury of distance would be one they couldn't afford. But for now they all seemed to savor this moment of personal space. Jay had slipped a yogurt into her backpack for breakfast, but she quickly realized that was useless. She'd have to take her mask off to eat, and you couldn't do that in here.

At the front of the room, another person with a covered-up face handed Jay her long white coat and new ID. Printed on the little piece of white-and-blue plastic were the improbable block letters: "Jay G., physician."

The residency program director and assistant director, both hulking six-foot-three men with baritone voices, came to greet the group. Everybody drew their chairs away from the wall and formed a misshapen circle. Now they were closer together, and the synapses between them seemed almost alive. This was what Covid had done. It had turned distance—or its absence—into an electric force.

"Welcome aboard," the program director said, his eyes flicking from masked face to masked face. "We're glad to have you here." He informed them that they would be functioning as interns, or first-year residents, a step up from the level of responsibility they had in their final medical school rotations. They would rotate units every two weeks, and each be given four or five patients. As they began reviewing the coming weeks of responsibility, Jay's mind was suddenly awash in all the questions that she'd been damming up. How much PPE would they get? Would they be able to put in their own orders, arranging for medications and procedures, or would their residents have to approve the decisions? How many Covid cases did the hospital have?

Jay's hospital had been designated as a Covid Center of Excellence, which meant it was accepting the bulk of coronavirus cases that came to emergency rooms within its hospital system rather than steering incoming patients to other hospitals. It was five weeks since positive swabs began surfacing in the city; more than ten thousand people had died in the tri-state area, and nearly a quarter million had been infected. Governor Cuomo had recently extended New York's stay-at-home order through at least mid-May, meaning nonessential businesses and schools would remain closed.

Jay had read the news each day with a sense of purpose that enabled her to mute the voice in her brain responsible for angst production, but that mute button was gone now. "We're going to go over allocation of PPE now," said the director. "Each of you will get one N95 per week."

Jay made eye contact with one of the other new interns. They were all familiar with the studies that said you should swap out your N95 every five to six hours to preserve its filtering mechanism. She resolved not to mention this N95 rule to her mom, at least for now.

"So how is everyone feeling?" the director continued. "Should we pick up your PPE?"

The group filed downstairs and outside, toward another area of the building designated for PPE distribution. Its main entrance was blocked off. Peering past the staff guarding it, Jay saw a long white refrigerated truck, which must have been filled with corpses. These trucks, kept at a chill 34 degrees and stationed throughout the city, could hold up to a hundred bodies. Some were marked by lettering on their sides that read, ominously: "Dead Inside."

Some of the interns had been fitted for N95s before. But there was a grimness to it now, a level of palpable fear in the slow, clear instructions they were given. Each of them had to put on a mask and hood, which was sprayed with aspartame. Then they had to make sure they couldn't taste it while doing a set of exercises. Jog in place, nod your head yes, shake your head no, say "Hello" and talk for thirty seconds. Jay received her one N95 for the week and was told to mark the side of it that was exposed, which would be important to track when she retrieved it for reuse. They were informed that the hospital had run out of foot and hair protectors, which was also added to Jay's file of data she wouldn't share with her mom.

The hospital had made various physical adjustments for the weeks of crisis care. The hallways were lined with red signs: "Do Not Enter without an N95" and "No Visitors." The hand-sanitizer dispensers were being frequently refilled, but still they ran out at least once a day. New workstations were erected in rooms that hadn't been used in years, where the floors and tables were coated in filmy layers of dust.

Before receiving her list of patients, Jay returned to the lobby to activate her badge. She waited in the hall by the receptionist's desk. As the line inched forward toward the security guard on duty, she studied the linoleum floor. The lobby felt foreboding in its uncharacteristic hush; it was all bleak gray and the smell of antiseptic.

When Jay got to the front, she introduced herself. "Hi, I'm a new . . ." She trailed off. She found that she literally couldn't form the syllables to say "physician." It felt too foreign. "I'm new."

The guard squinted at her, nodded listlessly. "We gotta take your photo," he told her. "Stand over there." He gestured toward a nearby wall. "Oh, but first take off your mask."

Jay froze. "Uh. What?"

"You gotta take off your mask for the photo," he said.

Jay's heart started hammering. Her panic was disproportionate to the situation, but it felt like some sort of test—a "gotcha!" moment to see if, hours after entering the hospital, she would casually remove her protective equipment. But the guard just stared at her expectantly, so she slipped off her mask and smiled for the camera.

As Jay stepped aside, waiting for her ID to be ready, the middle-aged man standing behind her in line moved forward. He swaggered toward the guard like they knew each other, ran a hand through his thinning white hair, and adjusted his black leather jacket. "I'm here for a body pickup," he said in a thick Irish accent.

Jay turned her head to catch the man's facial expression, wondering if he could be joking. He laughed at her look of alarm, and his bright blue eyes glimmered like they, too, were in on the joke. "This is my sixth one today," he said, waving a pink slip of paper in the air. "I'm a funeral director in the area. I'm used to it by now."

"Oh, okay," Jay said.

As she turned to head to the elevator, he called back to her. "Make sure you wear that thing." He pointed to her mask. "I don't want to have to come back here for you."

As young doctors do, Jay had envisioned the things that might go wrong with her first patient. A rapid response. An airway code. She could overlook a possible diagnosis, give too large a dose of medication or, alternatively, one too small. The margin for error might be nonexistent. And the stakes were no longer a grade on a med school exam; it was somebody's life.

But at least medical care is typically collaborative, a choreography set by other doctors, by the patient's family, and by the patients themselves. What Jay hadn't anticipated for her first assignment was how isolating it would feel. Her patient, Ms. Parker, was nonverbal and had no family contacts on file.

Ms. Parker's information was handed over to Jay unceremoniously. She sat down in the workroom and opened her electronic medical records to review the facts. The woman was fifty-four years old, had tested positive for the coronavirus, and had recently had a stroke. She was homeless.

Since Ms. Parker was nonverbal and asleep, the attending physician said there was no reason to visit the bedside right away. After all, the hospital was trying to conserve PPE. But Jay decided to pass by anyway and peek through the window. It would feel wrong not to lay eyes on the first-ever person placed in her care. The patient's body was covered by a heap of blankets, her face concealed by a mask, and her eyes closed. She was on a high-flow nasal cannula, a thin plastic tube threaded through her nose to deliver her oxygen.

The woman had received a cognitive examination when first admitted, which had revealed that she wouldn't be able to make her own decisions on care. For other patients like her, the doctors would call up her family. But Ms. Parker had no emergency contacts, nobody to weigh in on whether she should be resuscitated or intubated if her lungs or heart failed. She had no primary-care provider, and her electronic records had zero information on her medical history. Jay wished for a moment that she had some other first case—someone who could communicate with her the way she'd envisioned for a first patient, who validated her training instead of scrambling all the clinical norms. She couldn't even rely on many of the typical protocols for patients with shortness of breath and altered mental status, because it was still unclear what applied to Covid treatment. The guidelines kept shifting.

Jay learned that the hospital ethics committee had been consulted about Ms. Parker's case, which brought a measure of relief. There were other voices involved. The committee had determined that if her heart failed, she shouldn't be resuscitated. The odds were too slim that she could survive. Jay settled at her desk in the workroom to study the patient's chart. She noticed that Ms. Parker hadn't eaten since her admission a week ago. She'd been sustained only by sugar in her IV fluids. Jay wondered if they could give her IV nutrition, but the decision had to be routed through the ethics board. She called the board and reached an answering service, so she left a message and made a mental note to call back the next morning.

The day slipped by in a flurry of new faces and information. On the 6 train home that evening, Jay pictured her patient in that tangle of hospital linens. She would do anything in her power for Ms. Parker's comfort. After years of school, years of wondering whether she was qualified to be a real doctor, of *hoping* she was qualified to be a real doctor, Jay didn't want to lose her first patient. Besides, Ms. Parker was only fifty-four.

Late that night, around 3:00 a.m., Ms. Parker's lungs started to fail. One of the night nurses saw and paged the overnight team. Night float—the residents on overnight call—rushed over, but by the time they arrived, it was too late. When Jay came in the next morning, there was nothing for her to do.

"We've already filed for the death certificate," the overnight resident told her.

Normally the next step would be for Jay to call family and let them know Ms. Parker had passed. Here all she could do was take her own private moment to mourn.

The rational part of Jay knew there was nothing she could have done differently. But God. That didn't change the fact that she'd lost her first patient. Day two of her medical career, and there were no lives saved, just another added to the death toll. Had she missed something?

One of the residents must have noticed the look of alarm in Jay's eyes. "I looked at her chart too," he said quietly, tugging her to the side. "There was nothing you could have done differently."

The resident was doing a residency in psychiatry but had been redeployed for frontline Covid care, and he seemed attuned to the emotional register of the room. Perhaps due to his training, he knew the right words to counter a wave of panic, even if it was the doctor who was in crisis and not the patient.

Soon Ms. Parker's body would be wrapped up and taken to an overflowing city morgue. Jay wondered how many had died alone like this in the last month, people whose hearts stopped short in the middle of the night with no phone calls to place the next morning. The phone calls to family were impossibly heavy, but the knowledge that there were no calls to make brought its own kind of heartbreak.

Later that week, Jay was assigned another patient who had recently coded with a pulmonary embolism, or a clot in the lungs, most likely caused by Covid-19. This time the patient had survived the code. But she was younger than Jay, just twenty-three years old. Jay felt a visceral jolt of fear. The masks were still in short supply.

Around the time Jay first reported for hospital duty, the city was at four times its normal death rate. There was a New Yorker dying almost every two minutes. Jay's first day came just past the peak of city Covid deaths—but that was only clear in hindsight. At the time, she and her team wondered if the count might continue to mount. The uncertainty bred its own kind of nebulous angst.

It wasn't just rookies like Jay feeling the weight of the frontline care. Even the country's veteran doctors found themselves tested by the weeks of Covid crisis. There were doctors who served on the front lines of September 11, Hurricane Katrina, the AIDS crisis, the West Nile virus scare, the anthrax scare, Ebola, some combination of the above. But Covid felt worse—the fear heavier, the death toll higher.

The difficulty for frontline providers was partly in the lessons that had to be unlearned. Many of the mores of medicine were challenged, some cast off entirely. Dr. Fred Valentine, who helped treat Bellevue's first HIV patients, developed a lecture for his trainees on treating novel infectious diseases entitled "We've Never Seen This Before." He quickly realized that some of the lessons from the AIDS crisis held true during the Covid era: namely, when treating new diseases, sometimes old rules and traditions no longer applied. "Figure things out. You guys are smart," he told his trainees. "Particularly in the area of infectious disease, you're frequently seeing things no one has seen before. That means you gotta think. Put things together."

During the early weeks of Covid, one of the new challenges for providers was widespread concern about ventilator allocation. The country's hospitals were estimated to have about sixty-two thousand ventilators at the start of the pandemic, and the Strategic National Stockpile didn't have sufficient backups because of a contract with a private firm that fell apart in the years preceding the Covid outbreak. As the surge of Covid patients mounted, many doctors worried that ICUs would become overwhelmed, and there wouldn't be enough ventilators for everyone in need. That led to difficult questions. In a choice between two sick patients, who would get priority access to a ventilator? And how could this decision be made fairly? Some patients, and doctors, worried that racial bias could affect these choices, just as Black and Hispanic patients were getting harder hit by the disease.

In 2015 the New York State Task Force on Life and the Law (created during the AIDS crisis) worked with the New York State Department of Health to revise earlier guidelines on how to allocate ventilators during a pandemic. The task force decided that rather than giving out ventilators on a first-come, first-served basis, the decision had to be made by taking people's health conditions and likelihood of survival into account. The crux of the decision should be a person's sequential organ failure assessment (SOFA) score, which measures any dysfunction in six organs and systems (respiratory, coagulatory, liver, cardiovascular, renal, and neurologic). The task force stipulated that ventilator access should never be determined by nonmedical factors, such as race, ethnicity, sexual orientation, socioeconomic status, or ability to pay. It was important to be as objective as possible.

Still, these were guidelines, not hard-and-fast rules. Doctors, patients, and their families worried about what extraneous factors might mold these life-or-death decisions. If the surge kept mounting, if ICU beds got more overwhelmed, would someone uninsured, Black, or Hispanic be shunted aside? As early as March and April there had been cautionary tales of Black people in New York denied care, like beloved Brooklyn middle school teacher Rana Zoe Mungin, who died of Covid-19 after being twice denied a coronavirus test. The disparities were becoming rapidly apparent.

During their onboarding trainings, some of the new interns heard that there were hospitals whose proposed guidelines suggested that no one over seventy-five should get a ventilator if resources grew scarce; their survival odds were too slim. People didn't usually contend with such heavy considerations in their first months out of medical school, while they were still learning their way around the hospital halls and remembering to introduce themselves as "Doctor."

Another challenge for new doctors was the pervasive fear of infection. For the most part, neither senior physicians nor new graduates had been trained to worry for their lives while caring for patients. Tuberculosis patients carried the risk of infecting their doctors, and providers treating people with AIDS learned to draw blood carefully so as not to be infected by a needle stick. But for the most part, their concern was for the patients and not themselves.

Until the Covid weeks. Suddenly, every hospital worker regardless of seniority carried the same worries: Did I don and doff my PPE correctly? Is my mask snug enough to protect my airways? These questions had to come before patient care. In the first few days, there were moments when a cardiac patient coded, and overzealous doctors started to rush into the room without their gowns and shields. The more cautious physicians had to remind everyone to protect themselves first.

Affixing your own mask first flies in the face of what doctors are trained to do—leap into action for their patients. They are intimately familiar with the importance of speed. They can quantify all that's lost in oxygen and brain activity in taking even three extra minutes to properly secure PPE before responding to a code. So those steps for self-protection can feel like a selfish part of the medical equation, a subtraction from patient care.

For Jay, and many of the other early graduates, the hardest part of this was the wedge it drove between doctor and patient. After all their studies of patient-centered care, they were told they could barely spend time with patients. They had to minimize their minutes at the bedside. They struggled to see faces through the layers of PPE. Studies indicate that a medical worker's physical touch can promote better health outcomes, yet doctors had to be cautious about leaving themselves exposed.

This fear of infection made for a difficult training environment. But it also prompted doctors to think more creatively and intentionally about their communication with patients. They reflected on how to use their precious time by a patient's bed, or which words of comfort they could offer, since visitors were not able to be present. Physicians became hyperattentive to how they could use their body language to communicate, especially since patients who were hard of hearing couldn't fall back on reading lips. They were vigilant about standing with their feet open and arms uncrossed so patients would know they were listening and engaged. Some felt a new intimacy with their patients, who couldn't have relatives in the room. The connection was also deepened by the fact that patients knew, and appreciated, the risk their providers were taking to offer them care.

The challenges of treating Covid-19 patients didn't erase the need to build trust and connect personally—it just made this work more complicated. The unconventional thinking that novel diseases require can help trainees to become better doctors. "The new disease may not fit into the body of knowledge you already have," said AIDS expert Dr. Valentine. "You have to prepare to be surprised."

Jay's phone calls home to family were fraught. She was used to imparting the details of her day unpolished and uncensored—what she ate, what friends she saw, new plans percolating in her mind. It hadn't typically occurred to her to hide anything from her mom. Now she was sifting through her work hours to find the stories that wouldn't cause unnecessary stress. Nothing too heavy. It felt like tiptoeing on a creaky floorboard, hoping the weight of a sudden step wouldn't be cause for alarm.

Jay told her family the anecdotes that wouldn't make them nervous about her own possible Covid exposure. There was the woman with abdominal pains who thought she was short of breath because of Covid, but actually just had anxiety about returning home to an empty apartment. There was the man with diabetes who put aside his concerns about flying during the pandemic and traveled from New Mexico to New York for an emergency surgery, which saved his foot from amputation. There was the Chinese-speaking woman awaiting surgery whose interpreter mistakenly told the operating team she had eaten three meatballs, rather than three pills, delaying her procedure until the miscommunication was sorted out.

There were other incidents that Jay found herself on the verge of sharing—but they were too frightening, so they didn't pass her new filter.

One of those came in Jay's second week. She was working in the step-down unit, which some considered just as high-pressure as critical care. Step-down provided an intermediate level of care, in between the ICU and general wards. At least in the ICU the patients were on ventilators and other machines that kept their oxygen steady. In step-down, they could crash at any minute. Lots of them did.

On her first Monday in the step-down unit, Jay pulled up a female patient's chart before going to introduce herself.

This particular patient, an older African American woman named Ms. Harvey, had come to the hospital with a urinary tract infection in early May and been prescribed antibiotics that caused a rare and severe reaction. Days later, she was passing black stool. Then she swabbed positive for coronavirus. Between her gastrointestinal issues and her coronavirus test, the doctors decided to transfer her to the step-down unit for more oversight.

"Good morning, Ms. Harvey," Jay said. "I'll be taking care of you."

"Good to meet you," Ms. Harvey said. She was so petite that

her mask overtook most of her face, which was freckled, with delicate features.

"How are you feeling this morning?"

"I'm feeling okay," Ms. Harvey said.

"Your stool is still bloody, so I think we should start a blood transfusion," Jay explained.

Ms. Harvey nodded. There was something warm and amicable in the woman's manner that made Jay feel instantly tied to her.

An hour or so later, Jay was checking on her other patients when she was intercepted by one of the nurses in the hallway. The nurse looked panicked. Like Jay, she was an early graduate, on the job for just ten days.

The nurse held out a bloodied bedsheet. "This is from your patient," she said, gesturing toward the soaked white linen. "I just pulled it out from under her. She won't stop bleeding."

Jay broke into a run toward Ms. Harvey's room, with the nurse close behind her. Ms. Harvey, who had started her blood transfusion, was now bleeding out onto the bed surrounding her. Her brows furrowed as she looked back and forth from Jay to the nurse, waiting for some kind of reassurance.

Jay went on autopilot. "Get her vitals again," she said. She ran to a computer and ordered a CT Angiography scan of her abdomen and pelvis, which she had only ever read about, never seen. Her resident and attending were dealing with a coding Covid patient, so Jay was on her own for this one.

Jay called downstairs to gastroenterology and radiology, then returned to the patient. "We're going to bring you down for a scan," she told Ms. Harvey. She figured they would keep the blood transfusion going as they moved.

The transporter showed up a few minutes later. Jay could sense that Ms. Harvey was nervous, so she walked alongside her to the elevator. "I'll come down with you," Jay said, and Ms. Harvey eyed her with wordless gratitude. This wasn't common, but

Jay didn't feel like she could let the woman go down alone; her anxiety was palpable. Jay hit the button for the basement, where imaging was located, and they rode down together in silence. The elevator was small, with barely room enough to fit the bed, the transporter, and Jay.

Then suddenly they felt the elevator buckle and lurch. Jay hit the Door Open button, but there was no response. She hit the button repeatedly, wildly. Nothing. They were stuck. Just her, a Covid-19 patient in the midst of a blood transfusion, and their transporter, who, given his role, wouldn't have clinical training.

Ms. Harvey turned her eyes upward, locking in with Jay's. "Hey," she said. "Do you know what to do if something happens to me right now?"

Jay felt her heart hammering against her sternum. "Yeah, I do," she said, mustering any iota of confidence she could find in her body. "Of course I do."

Do you know what to do if something happens to me? The question thumped against the insides of Jay's brain. She looked at the tiny woman on the bed. Her eyes were closed. Blood continued its steady drip into her arm through the IV. Did Jay know what to do? Was that any help? In that moment, technically knowing what to do and being able to do it seemed vastly different. Worse still, her resources were limited to whatever she could do in a stalled elevator.

Jay had been trained in advanced cardiovascular life support, but her training had been virtual—a few hours in her bedroom clicking through modules on her laptop. Typically ACLS training required in-person certification, but, for Covid reasons, that was waived for the early graduates. At the time, the simulation seemed far afield from a real-life medical emergency. Now it felt even further, like a game of *The Sims*.

Jay tried to remind herself that years of schooling had prepared her for this type of situation, and there were gatekeeping powers-that-be who had deemed her capable of handling them.

She had answers, classroom lessons tucked away in the recesses of her mind.

Jay began to imagine that this was a scenario on an exam. Let's say the woman's mental status has changed because of low blood volume. She would then check Ms. Harvey's pulse and elevate her legs. She would keep the transfusion running, careful not to let the patient's saline levels drop. Okay, let's say her heart stopped. Jay would do chest compressions. Fifteen of them for every two breaths. She pictured herself with her hands on Ms. Harvey's chest, pressing down forcefully. (Then she thought about the requisite motion involved in that compression. She would be leaning over Ms. Harvey's body with her mouth close to the woman's face, which meant near certain coronavirus exposure. She tried to quiet that thought the moment it bubbled up. At least she was wearing her N95.)

The transporter desperately tried to find a signal on his cell phone to call for help. The minutes dragged on. After eleven of them, agonizingly slow, the elevator doors abruptly rattled open. Someone on the other side had called for the elevator, and apparently somehow prompted its response. Jay stepped out into the fluorescent-lit hallway, which now seemed like a mirage. The band of three—Jay, patient, transporter—proceeded down the hallway, too rattled to speak.

Outside the imaging room, Ms. Harvey looked up at Jay. "Will you come inside with me?"

"I'll be right out here the whole time," Jay replied.

Back pressed against the wall, Jay tried to take a calming breath. She felt suddenly like she had passed this unanticipated, hypothetical exam. Somehow, despite all her reasons for self-doubt, she had managed to gain her patient's trust.

Doctors who want to connect deeply to their patients must confront a reality of the American health care system: many minority patients, especially Black people, have long distrusted the country's medical providers, knowing the field's history of racial prejudice and outright abuse.

In 2018, two researchers, Dr. Marcella Alsan and Marianne Wanamaker, made a profound, though predictable discovery: African American men living close to Macon County, Alabama, were significantly less likely to visit physicians after the year 1972. That was the year a government-led study came to light, whose findings ultimately told us more about racism than about infectious disease: the Tuskegee Study of Untreated Syphilis in the Negro Male. Starting in 1932, the country's Public Health Service, a branch of the Centers for Disease Control and Prevention, tracked roughly 600 low-income Black men in Tuskegee, Alabama, 399 of whom had syphilis. Many of these men were sharecroppers with no prior experience visiting a doctor. The men were told that they would be treated for "bad blood." But they weren't given the preferred drug for syphilis, penicillin, which became standard in 1947. Instead, their diagnosis was kept a secret as the researchers examined the disease's effects on them. As a result of not receiving proven care for a treatable condition, the men suffered. Some went blind; others deteriorated cognitively. Many of them died. Some spread it to their partners and children at birth. This went on covertly for four decades.

The Associated Press broke the story in 1972, prompting public outrage and forcing the study to shut down. But the aftereffects continue to reverberate. Decades later, Dr. Alsan and her fellow researchers still meet African American people who say their parents drove many miles away to take them to nonwhite pediatricians when they were young.

The Tuskegee study is the most infamous example of America's medical racism in the last century, and knowing its cruelties is critical in forming at least an elementary understanding of the Black community's lack of trust in doctors. But it is only a single chapter in a long history, too often treated as exceptional instead of as a troubling norm. "It is a mistake to attribute African Americans' medical reluctance to simple fear generated by the Tuskegee Syphilis Study, because this study is not an aberration," writes ethicist Harriet A. Washington in *Medical Apartheid*. In reality, medicine was never free of the symptoms of racial hatred pervading all other American fields. Nor is it today, Washington adds: "The greatest tragedy of the study is that it has failed to serve as a cautionary tale."

Medical researchers have long subjected Black patients to the riskiest studies, particularly those testing nontherapeutic drugs. In the 1920s and '30s, researchers under the auspices of the Rockefeller Foundation deliberately infected 470 Black syphilis patients with a deadly strain of malaria, killing some of them. The researchers then disguised their cause of death. Meanwhile, across the country, a number of states started sterilization programs in an effort to control minority populations; the "Mississippi appendectomies" were unnecessary hysterectomies performed on Black women at southern teaching hospitals, often without their knowledge. A third of the patients were under the age of eighteen. In *Medical Apartheid*, Washington points out that the sterilization of Black women, taking place just a century after slavery, continued the legacy of white domination over Black women's reproduction: "During slavery, black women had been forced to procreate, but now they were being forced into sterility," she writes. "The consistent factor was white control."

For decades, Black patients remained more likely than white patients to receive undesirable procedures like hysterectomies and amputations, even when less invasive options were available. Simultaneously, they became less likely to receive important operations like hip replacements and surgeries for cancer. Medical journals disparaged Black people as "disease-spreader[s]" and as patients "unwieldy, unwilling, unsatisfactory" in their adherence to medical treatments, language that served to justify widening health disparities. Physicians predicted that syphilis or tuberculosis would bring an end to the Black population.

The story of Henrietta Lacks, a poor Black woman from Baltimore who died in 1951, encapsulates the medical field's dehumanizing treatment of Black patients. Cancerous cells were taken from her body without her knowledge before she died and then used in experiments key to modern medicine for generations, with no benefit to her family. After her death, scientists went on to publicize her name and even publish her genome without asking her family for their consent.

The dehumanizing treatment that Henrietta Lacks experienced isn't a phenomenon of long ago. In 2001, three white medical students at the University of Alabama at Birmingham were exposed for wearing blackface at a party, costumed as Stevie Wonder, a character from the cartoon show *Fat Albert*, and a generic Black woman. The photos, leaked to the media, raised the question: If they could caricature Black bodies for Halloween entertainment, how were they treating Black people's bodies in the examination room?

More recently Black patients have used Twitter and Instagram to share advice on getting quality care from an unequal system, including documenting when white doctors refuse a test that's been explicitly requested, extending a tradition that began with advice columns in *Ebony* and *Essence* during the heyday of Black print media. African American patients continue to receive lower-quality services across numerous categories, including HIV, prenatal care, and preventive care. Black mothers are more than two times as likely to die from pregnancy-related causes. For many Black patients, being able to get care from Black doctors is essential. But it's also immensely difficult, because Black people have had so few opportunities to enter the medical field. This access problem was compounded early in the twentieth century by the

Flexner reforms, which had closed down the majority of Black medical colleges.

The distrust of doctors that runs deep in minority communities becomes even more of a challenge amid the movement for patient-centered care. Care based on personal relationships—the kind that drew Jay, Gabriela, Iris, Sam, and others to the profession—isn't possible without trust. This is a trust that goes beyond "Is this going to hurt?" It requires vulnerability from both patient and provider—and it comes with life-or-death stakes. When a patient does not believe their provider will give them proper care, they might walk out of a clinic or emergency room even when they desperately need treatment. Even if they are on the verge of kidney failure, or cardiac arrest. If someone feels ignored or exploited by their physician, they won't seek out care. The cost of that decision could be their life. At a population level, the consequences are staggering. As the cardiologist Dr. Richard Allen Williams put it some thirty years ago: "A person with chest pains may be so angry at the medical system that he may refuse to go to the hospital and may die at home. If such behavior occurs on a large scale, the effect that it will have on morbidity and mortality statistics is obvious."

But you can't force a relationship with a population that has historically been betrayed by its doctors, and hasn't often seen itself represented in medical ranks. Dr. Louis Penner, a researcher who studies health inequities, remembered having a conversation with a physician friend who was an avid proponent of emerging, less paternalistic styles of care in the early 2000s. "There were all these predictions of what we would find when doctors engaged in patient-centered care," he said. "But it turned out the patients didn't act the way they should have. Because they didn't trust the damn doctors!"

Histories of medical racism are rarely confronted head-on in the medical curriculum, further deepening mistrust in the field.

"I often joke I want my tuition money back," said Dr. Uché Blackstock, the founder of Advancing Health Equity. "Because there's so much I didn't learn. They never told us that Henrietta Lacks's cells were from a Black woman from an impoverished neighborhood. That's a story that Black physicians now know well, but at the time we weren't taught it. Medical schools need to address the historical context of racial injustice in medicine."

In recent decades, research has pointed toward a mechanism for countering some medical distrust: Train more Black doctors. Train more Spanish-speaking doctors, more queer doctors, and others whose shared sense of identity with patients can strengthen bonds and improve medical care. This field of study, on "concordance," examines whether patients and doctors have better relationships when they share aspects of culture and identity, including race.

Several years ago a group of researchers set out to determine whether Black men who visited Black doctors might see specific improvements in their health outcomes. The researchers recruited thirteen hundred Black men during visits to barbershops and flea markets around Oakland, California. These men, of different ages, incomes, and educational backgrounds, were asked to complete a survey about health care, and then given a coupon for a free health screening at a local clinic. The researchers created a freestanding clinic and staffed it with fourteen doctors, six black and eight non-Black.

Before their appointments, the men were given the chance to select what preventive care services they might want from the doctors. The researchers found that once at the appointments, the men who met with Black doctors were more likely to agree to preventive services and more likely to receive additional elective tests beyond the ones for which they had previously signed up. They were more open to diabetes screenings, BMI assessments, and cholesterol tests.

The researchers found that the Black men visiting non-Black doctors turned down flu shot offers more than those visiting

Black doctors, even when offered $5 or $10 incentives. The differences were most pronounced in the patients who indicated a general mistrust of the medical system in their pre-appointment surveys.

The researchers wanted to know why the disparities between those who saw Black and non-Black doctors were so pronounced. They looked over doctor notes, patient feedback, and outside survey data, and found that Black patients were more likely to raise personal issues relating to their health with Black doctors. They were also more likely to believe their doctors understood them, and made them feel at ease.

The Oakland Men's Health Disparities Project was fairly small and focused only on Black men. But recent papers have continued to build on its findings. One study, published in 2020, examined 1.8 million births in the state of Florida between 1992 and 2015 and suggested that Black babies are more likely to survive when birthed by Black doctors. Other research has found that when minority patients see minority doctors, their visits are longer and rated more satisfactory. When Spanish-speaking patients see doctors who speak their language, they are more likely to adhere to their prescribed medications. And one study found that Black patients who see Black physicians are more likely to accept the doctor's assessment of their risk for getting lung cancer than when they see a non-Black doctor. There is also strong evidence that physicians' implicit biases and stereotypes can influence their care, contributing to health disparities.

This does not suggest that people should only be treated by doctors of their own race. But it suggests the need for doctors of all different backgrounds in hospitals and examination rooms. "Absent some wonderful societal shift, it's a health imperative that people have the possibility to see concordant doctors," said Dr. Susan Persky, an associate investigator at the National Institutes of Health. "This bolsters the call to make sure more students of color get into medical school."

Researchers have also probed other kinds of concordance beyond race. First, the gender matchup can matter. A 2018 study published by the National Academy of Sciences found that female heart attack patients have a higher likelihood of surviving when treated by female physicians. Another found that when a doctor and patient's gender don't match, the patient is less likely to consent to screenings for breast, cervical, and colorectal cancer.

Sexual orientation affects medical care too. A study of lesbian, gay, bisexual, and transgender people in New York conducted several years ago found that 40 percent felt there weren't enough medical professionals trained to provide them with care. Research has shown that straight physicians sometimes exhibit implicit bias toward their gay and lesbian patients, and their biases can impact the medical decisions they make. Patients pick up on that, and it adds to the mistrust bred by history. A clinical psychologist surveying almost one thousand medical students found that some who identified as heterosexual were reluctant to prescribe antiretroviral medications to their gay male patients for fear it would encourage unsafe sexual behavior.

Doctors are not just the sum of their technical skills. When they show up at a patient's bedside, their experiences and perspectives and identities inform their care, and how it's received. This is all the more true when they're treating patients whose stories mirror their own in some way. And it's especially true for doctors like Gabriela, Sam, Jay, Elana, Ben, and Iris, who want to form emotional bonds with their patients. They see many of their patients through the lens of their own personal histories—and their patients see them through their own lenses too.

Ten

GABRIELA, April 2020

Gabriela, masked and in scrubs, entered her first medical order into the NYU system: it was for Tylenol. A sleek capsule that she had popped into her own mouth innumerable times, but the formality of typing it out as a medical order was unnerving. She read over the dosage twice and then had it checked by a resident, who met the request with a good-natured smile.

For her first day, Gabriela was wearing a bright pink mask dotted with cartoon characters. "Can you tell I'm going into pediatrics?" she joked to one of the nurses.

She was assigned four patients, and three of them were ready for discharge that day. This felt strange to Gabriela, with her caretaker instincts. It was angst-inducing to bid her first patients goodbye the minute they became her responsibility, especially when they were scared to leave the hospital. Their eyes searched hers suspiciously.

"Are you sure I'm ready to go?" one older patient asked. She was a large woman, African-American, with short hair. To get ready for discharge she had changed from her hospital gown into a blouse and cotton skirt.

In the pre-Covid time, a discharge felt more celebratory. "I hope I don't get to see you again," Gabriela had liked to joke to her patients on rotations. She could get attached, especially to those patients who reciprocated her affections, but she was always happy to see them headed back to home-cooked food instead of applesauce and the limp colorlessness of hospital provisions.

But for some patients, leaving the medicine wards in the pandemic era meant a fog of uncertainty.

"So am I immune now?" the woman asked.

The flood of questions continued from there, all well-reasoned. *Is it possible I could expose my family? What do I do if my oxygen levels drop again? Do I still need to isolate?* For many patients, the Covid symptoms had come on abruptly—what started with a cough suddenly became an inability to breathe—so they worried about declining again just as rapidly.

"We wouldn't let you go if we didn't think you were ready to leave," Gabriela told her patient that morning, planting a degree of certainty into her voice. This was the verbal gymnastics required of her, when she introduced herself as the new intern and then assured these patients that she was confident that they were healthy enough to depart. Not only was Gabriela new but so was the virus; so, too, were the circumstances. There was still so much the doctors didn't understand about the body's response to Covid, including how immunity after infection might operate.

The patient grabbed Gabriela's hand and gave it a circulation-constricting squeeze.

"I know it's scary," Gabriela said. "I know." For a moment, feeling her palm on the patient's, she thought of Grammy, who must have squeezed so many hands and seen so many searching eyes. *For you, it's a normal day. For the patient, it's either the best or worst day of their life.*

Gabriela could have sat there for an hour. But her attending physician had given her strict instructions not to spend too long at her patients' bedsides. They were doing everything they could to minimize Covid exposure. That meant prioritizing safety, not emotional comfort. So eventually Gabriela had to disentangle her gloved hand from the woman's and gently remind her that it was almost time to leave.

It was something of an identity crisis for Gabriela. She had sketched an outline of the type of doctor she wanted to be. The doctor who would play liquid nitrogen tricks to distract a young girl getting her warts frozen off. The doctor who might find out a patient's favorite sports team and stop by their room with the latest score (especially if that team was the Patriots). That doctor would have Grammy's alchemy of tenderness and resolve. Once Gabriela saw an intern run out to buy a Pepsi Zero for a patient. That right there, she thought, that'll be me.

But that role wasn't entirely possible for these weeks—not with all the layers of protective plastic, and with distance from the beds strictly enforced. She was in the Covid wards to help while her team was overrun, a pinch-hitter of sorts. She moved rapidly between her patients, willing herself not to linger, grasping for the right words of comfort before moving on to the next.

Occasionally, in those first days, Gabriela's patients mixed her up with the nurses. They weren't used to having a Hispanic woman as their doctor. Many of the Black and Hispanic faces they saw in the hospital were food service workers, technicians, or custodial staff.

She could feel them sizing her up. "I'm the *intern* on your team," she repeated. "I'll be taking care of you."

Introducing herself as a doctor still felt new. But she had experience with patients who doubted her medical credentials from her rotations as a student.

The script was familiar from those days. A patient would look over to the white attending physician with deference. Then, when Gabriela spoke, they would stare in confusion and ask something like: "Are you my nurse?" Gabriela would swallow hard, forcing down indignation and choking up etiquette. "Nope," she'd say. "I'm a medical student."

Once, during medical school, Gabriela was rounding with her clerkship team, and the senior doctor, who was white, introduced her to a patient as "mi amiga."

"This is mi amiga, Gabriela," the attending physician said to the patient, a smile dancing on her lips.

Gabriela's face flushed. Could she have misheard? There was no way. She tried to muster a response—now the patient's eyes were on her, and the other medical students'—but all she wanted was to be somewhere else. She couldn't react. This physician would be giving her an evaluation at the end of the week. She didn't want to be the trainee who stuck out as overly sensitive or fragile. *Mi amiga.*

"Nice to meet you," Gabriela said to the group. Her pulse was racing, but she decided she must be overthinking. None of the other students seemed to have registered the comment as offensive, or at least none of them had offered a reaction. Maybe this was just a joke to be taken in stride.

Later, after the group disbanded, Gabriela's friend pulled her aside. "That was so uncomfortable," she said. "I was so offended, I can't imagine how you must've felt. I'm going to report it."

"I hadn't thought about reporting it," Gabriela said.

"You shouldn't have been referred to that way," her friend insisted.

Gabriela hadn't realized that she'd noticed. It brought some marginal relief to know it wasn't all in her head.

Gabriela felt like the correspondent from hell when she called home. She tried to excise the bleakest hospital stories, as if they were growths that she could just snip. Her family didn't need them. They had their own sources of stress. "How long do you think the salon will have to stay closed?" her mom wondered aloud on the phone.

None of them could be sure. Her mom went back to the salon

once a week to pay bills, and she cried every time. She couldn't handle seeing it empty. Some of her clients were calling to ask if she'd send stylists to their houses, but she told them absolutely not.

One afternoon she spent three hours trying to apply for the government's Paycheck Protection Program, a loan to help cover payroll. She got her application materials together, and then the website crashed.

"This is so hard," she told Gabriela, in a voice like her daughter's, carried on a gust of purpose. It wasn't just the stress of paying the workers and bills. Her mom's identity was tied up with the business. Every piece of the place—its interior design, its staff, its reputation—was formed by thousands of hours over more than a decade, early mornings spent poring over paperwork and late nights sweeping up hair. Phone calls to a stylist about to give birth, a discount for a customer whose dad had just died. The salon was built on every ounce of her. It was the far-fetched vision board of a girl from Springfield's inner city.

Gabriela's mom had only ever shut down the salon once, for a week, when she was relocating to a different space, a retro salon from the 1950s that she had gutted and redone. The whole family went to move supplies in the middle of the night, traipsing back and forth with their arms full of combs and hair dryers until three in the morning.

Growing up in Gabriela's home meant understanding something fundamental: the best sort of job is more than a job. It isn't about clocking in nine to five, or nine to nine. It's so deep that it becomes part of your body. Every event you see through the lens of the people you serve. Weddings mean fishtail braids; funerals mean a sleek, subdued blowout. Balayage for the first day of school. Clients diagnosed with cancer came in to have their heads shaved when the salon was closed, for privacy. Old women came in just before they died, wearing kitten heels to feel glamorous in their last days. There was a certain look that Gabriela's mom got in her eyes when she talked about it. It was the same one Grammy had fixed on her face when she spoke about her patients. It was a glimmer of weighty pride.

Now the salon's future was uncertain, like that of so many small businesses across the country; more than a hundred thousand would shut down permanently by May. Sometimes Gabriela felt guilty talking about her worries at the hospital, because at least that stress meant the job was still intact.

Gabriela's stepdad Tommy was still working, but his job also bore new risks. He and his son, who they called Lil' Tom even though he was thirty-four, owned a delivery route that brought Thomas' English Muffins to wholesalers in the area. Tommy woke before dawn each morning to make his deliveries all along I-90.

"Are you being safe and wearing a mask?" Gabriela asked him over FaceTime. Tommy was diabetic, which put him at extra risk if he got Covid. He was now waking up even earlier than normal to avoid other delivery people on his route. Normally he would leave at 4:00 a.m., but now he left at 2:00, dragging himself out of bed and onto the darkened highways before the surrounding towns began to shake themselves awake.

Gabriela felt a shift in her family dynamics when they worried aloud. Normally she was the one who was being fretted over. But now her mom and stepdad were coming to her for advice, asking her how to keep safe as they choreographed their new reality. It might have made her feel adult, if it weren't also so unsettling.

BEN, April 2020

At Montefiore Hospital, the early graduates were called the Coalition Forces. This made them sound like a group of troops called to gun down adversarial forces. Meanwhile, they were just scrambling to help where they could, and not to misstep. It did feel like there was a war raging, sort of. The Covid death toll was climbing, and Montefiore was under siege. But to call the medical students graduating on Friday and clocking in Monday morning the Coalition Forces seemed a bit overblown.

Ben heard his alarm ring just past 7:15 on the morning of April 20. He used the Sleep Cycle app, which tracked his motions at night and woke him in lighter phases of sleep, maximizing his REM rest. He was surprised by how normal he felt as he climbed out of bed and fixed himself breakfast and tea (never coffee), waiting for his roommate Sean, who was also starting in the Covid wards. There was this repellent force, like magnets, between the lofty terms that had conjured this day—*heroes*, *warfare*—and the utter banality of toothpaste and bagel crumbs.

The two walked from their student housing to Montefiore, where they would have their orientation, a few blocks down Eastchester. It was smaller than the site where Iris was, with just 396 beds. But the hospital was still an imposing structure, a gray fortress whose manicured front seemed out of place among the auto dealerships and delis around it. At its front sat a stone wall with silver lettering spelling out the hospital's name, and around it were clusters of plants in muted purple, green, and yellow that were always freshly trimmed. It gave off an aura of order and gravity, not warmth.

The surrounding neighborhood, Morris Park, was known a century ago for its large Orthodox Jewish population but now it was predominantly Italian and Hispanic. Its streets felt almost suburban in their residential sprawl, lined with vinyl-sided houses and mom-and-pop shops, with not much in the way of nightlife unless you counted the retro bowling alley one mile from the hospital where Will Smith filmed *Men in Black 3*.

That first day, Ben and Sean paused to take a photo outside the sign that read Albert Einstein College of Medicine. Any other year they might have posed for ceremonial pictures when they were all dressed up for graduation, snapshots for their families to hang on fridges at home. But with graduation on Zoom, they would need to make do with this prework photo in scrubs. Their celebration had been a low-key pizza and Pepsi affair.

It occurred to Ben, at some point that morning, that he had not yet said the Hippocratic oath. Einstein's virtual graduation ceremony had mostly focused on logistics for the students-turned-frontline-doctors. It was all of an hour, squeezed between their other online training modules (though they'd have a longer, more formal livestream version in May). Now Ben mulled over the words most associated with that oath: *Primum non nocere*, "First do no harm." The phrase wasn't included in the original Hippocratic oath, attributed to the Greek physician Hippocrates, which was written around 400 BCE. The original draft had other instructions: "Give no deadly medicine to any one if asked" and "Abstain from every voluntary act of mischief and corruption." But at some point a line that Hippocrates had written in some other text had been folded in: "The physician must . . . have two special objects in view with regard to disease, namely to do good or to do no harm."

The mandate to "do no harm" made the Coalition Forces labeling seem even more peculiar. Most of the time, military troops surely did commit some harm—that's their job. But that wasn't the nature of the Covid army. Ben wasn't there to fight back against the virus. Only his patients could do that. There was no magical cure, no ammunition but the strength of a person's immune system. He and his classmates were needed to help their patients and hospital system withstand the onslaught of the disease. In some cases, Ben's role might simply be to help his patients die with comfort and dignity. Orchestrating the right kind of deaths hardly seemed like winning a battle—but this was a strange sort of war.

Ben learned that his placement for his weeks on the front line would be in cardiac telemetry. This was the unit that absorbed some of the hospital's sickest patients, those who needed constant monitoring. They were hooked up to machines that displayed their heart rate and blood pressure at all times. What this meant for Ben was that his Covid patients would be some of the more severe cases—the patients who were dying, coding, lungs failing. He and the other early graduates at Montefiore, including Iris and Elana, would be acting essentially as subinterns, which meant they would still need sign-off from superiors before putting in medical orders. Still, they would carry all their own patients and work regular intern hours.

Some of Ben's classmates might have been intimidated by the assignment, but Ben had grown fascinated by end-of-life care, and more specifically the ways people respond when their family members near death. The interest started during a rotation he did at a New Rochelle hospital in his third year. The hospital was surrounded by nursing homes, so Ben's days became a string of conversations about end-of-life care. Mostly they revolved around patients' decisions on code status: whether they wanted to be marked as do-not-resuscitate (DNR) or do-not-intubate (DNI).

Up until Ben's New Rochelle rotation, his understanding of DNR and DNI orders had been, essentially: Do you want us to do everything we can to save the patient? Ben's area of focus, after all, was emergency medicine. The ethos was "intervene, intervene, intervene." No matter the ailment a patient presented, gunshot wound or appendicitis, the team jumped to fix it.

But working with palliative-care doctors, he learned that their approach was different. There were interventions that could stave off imminent death, but it wasn't clear whether they would help the patients meaningfully recover. You could prolong their lives by three months or maybe six, but often the care was invasive, uncomfortable, and costly. The question that patients had to tackle, with their doctors and relatives, was how to balance a wish to prolong their lives with a desire to die as comfortably as possible. An aging woman with multiple comorbidities deciding on a DNR didn't mean she was giving up on life. It meant she realized that if her heart stopped, if she coded, chest compressions would be traumatic and might break her ribs. A DNI meant a recognition that if her breathing stopped, and she had to be intubated, her prospects of survival would be slim.

This was the script: a relative, often a daughter or son, would come to the hospital, and the palliative-care doctors asked the family member the sort of questions that nudged them to a fair, well-informed decision. Did they understand how sick their elderly mother was? How did she define quality of life? Did she want to push for recovery so long as she'd be able to keep doing simple activities like watching TV and eating ice cream? Or did she only want to live so long as she could dress herself and walk to the grocery store?

The crux of these conversations was the question of how this patient defined a life worth living. It was philosophical, which piqued Ben's interest in a cerebral way. But it was delicate, too, almost impossibly so. Everyone knows vaguely that death is real and inevitable—until it comes close enough to congeal into questions of logistics. Then people will do anything to avoid confronting its reality.

For most of human history, death was less of a process than an event. Geriatrician Joanne Lynn likened it to bad weather; you would spot the storm on the horizon and brace yourself, but there was little to do except hope for the best. Illness was rarely a years-long protracted battle. Mozart died from what was likely strep throat. The silent-movie star Fred Thomson stepped on a nail and got tetanus. Emily Brontë caught a cold at her brother's funeral, refused to see a doctor, and died of tuberculosis four months later at age thirty.

Of course, things are different now. The age of antibiotics began in 1928, with the discovery of penicillin. During World War II the army, while studying chemicals for use in war, found that a compound called nitrogen mustard could be used for protective purposes too. This included fighting cancer of the lymph nodes, and the era of chemotherapy began. In the last two centuries, the average American's life expectancy has doubled. The share of the American population dying at home went from more than half in 1945 to just 17 percent in the 1980s. Today the number has climbed back up to 30 percent. Still, we've institutionalized death, moved it from the home to the hospital and from the eyes of family to those of the nursing staff, obscuring it from a public who'd rather not confront it head-on.

As medicine's capabilities have grown more sophisticated, its objectives have become more complex. Increasingly, doctors recognize that aggressive life-sustaining measures aren't always suited to a patient's needs or desires. That's the dilemma that Dr. Atul Gawande explores in his book *Being Mortal*, published in 2014. Dr. Gawande made the case that modern machines and medications have lengthened our lives, but they haven't taught us how to navigate the thorny question of when life should no longer be lengthened. As a result, our health care system spends much of its money on the last few months of life. We invest in expensive surgeries and drugs even when arteries turn so tough they crunch, and brains shrink so much that they knock around inside a person's aging head.

It was only in the 1990s that a nationwide wave of regulations emerged allowing for patients to be marked do-not-resuscitate, with forty-two states creating such protocols by 1999. Providers saw how initiating the delicate discussion on goals of care before someone arrives in an ICU can be a boon for both patients and their providers. In 1991 medical leaders in the city of La Crosse, Wisconsin, mounted a campaign asking medical workers to preemptively have end-of-life care conversations with their patients at nursing homes and hospitals across the area. Five years later, 85 percent of the city's residents who died had instructions written up before their deaths, up from 15 percent before the campaign.

Making these decisions preemptively is entirely within the realm of possibility, so long as it's prioritized. Medicare started reimbursing physicians for counseling on advance-care planning in 2016. Physicians say that more and more they're seeing patients familiar with the language that was so alien just decades ago, coming to the hospital equipped with designated medical decision-makers and living wills.

As in so many other areas of medicine, racial and ethnic disparities in end-of-life care have become increasingly apparent. A Kaiser Family Foundation survey of older adults in the United States with serious illness found that 65 percent of white adults had documented their preferences for medical care, while only 38 percent of Black adults and 41 percent of Hispanic adults had done so. Studies suggest that physicians tend to have longer goals-of-care conversations with white patients. Black and Hispanic patients are less likely than white patients to use hospice services or forgo life-sustaining treatments. Nonwhite patients are also more likely to say that talking about death with their providers might bring it closer to reality. This isn't surprising in light of medical mistrust in minority communities. It takes an immense amount of trust for a patient to engage their provider on how they want to die, and to believe their doctors when they say a course of treatment isn't likely to save their life.

Conversations on goals of care demand a delicate dance on the part of the physician, who must let patients make choices while still helping them think through relevant considerations. A doctor might be tempted to offer the medical information and then step back to let a patient and family decide on the best course of care, but that isn't always in the patient's best interests. Sometimes a patient needs to be told why a certain treatment plan might be overly optimistic.

In New Rochelle, Ben began to find the conversations with elderly patients and their families oddly soothing. More important than the script was the code of understanding. The patients and their families needed to grasp the medical stakes of their decisions—the likelihood of surviving intubation, the trauma involved in resuscitation. And after they got past their initial distress, there was sometimes a wave of gratitude. It was a privilege, they realized, to have the time and space to probe what they really wanted from their last months or days of life. The conversations were heavy, but they could also be meaningful and tender.

That semester, Ben was taught how the AIDS epidemic changed public attitudes toward palliative care. When HIV swept American cities, the medical system confronted a surge of young people who became suddenly, debilitatingly sick. Sometimes their partners realized the need for thorny conversations about the way these patients wanted to die. And unlike some of the country's aging population, many patients with AIDS had the mental faculties to be part of these discussions. The hospitals saw waves of young people wanting to die on their own terms.

As he started his Covid assignment, Ben wondered what end-of-life conversations would emerge from the coronavirus pandemic. Here was another mass of people whose health was suddenly deteriorating, some with the mental awareness to determine how they wanted to die. And Ben's role, with the Coalition Forces, was to try and minimize those casualties.

On Ben's first day of work at Montefiore, New York State saw 478 coronavirus deaths. Governor Andrew Cuomo was meeting with President Donald Trump to hammer him on the lack of available testing, and the state's largest nursing union was suing it for unsafe working conditions during the pandemic. New York Life Insurance and Cigna prepared to announce a $100 million fund for the families of health care workers killed while treating Covid-19 patients.

Ben kept thinking about those two holy letters now appended to his name, MD. It was all beginning to feel real.

The interns were divided into teams by color. Ben's was purple for cardiac telemetry, eighth floor south. It wasn't like Ben had some granular vision of exactly what his first day of work might look like, but he'd always had it in his mind that there would be some baby-steps introductions: How do you put in an order? Where's the bathroom? But within an hour of Ben's arrival on the floor, a patient coded. Ben watched from the doorway as two residents administered chest compressions. After less than an hour, the patient was declared dead.

During Ben's first week he was assigned three newly admitted patients, who seemed like straightforward cases. One had Covid but was relatively stable, and another had experienced a series of arrhythmias, or abnormal heartbeats. The third, Mr. Moore, had suffered a heart attack overnight. Nothing novel; Ben had seen a handful throughout his rotations in the emergency department.

Ben knocked on Mr. Moore's door to announce his presence. This was his pre-round, a chance to examine his patients before he presented updates to his residents and attending physician during rounds later in the morning.

"Hey, I'm Ben," he said to Mr. Moore, a hefty white man. "I'll be taking care of you."

When Ben came back to check on him later that afternoon, Mr. Moore's vital signs were normal. "Are you feeling any chest pains right now?" Ben asked.

"Nope." The reply was surprisingly cheerful.

"Would you like to speak with your family?" Ben asked.

Ben put Mr. Moore on the phone with his daughter. Listening to these types of exchanges brought equal parts pride and pressure. This was why Ben wanted to be on these floors, to get people back to their families. He could tell that his non-Covid patients, like Mr. Moore, were especially anxious to get out of the hospital, knowing all the cases of coronavirus that surrounded them in the building.

Forty-five minutes later, Ben went to pick up dinner with his senior resident. A local pizza place had sent free pies for the front-line workers. As he chewed, the PA system crackled on: "Rapid response, eighth floor."

"I'll take care of it," the resident said. "You keep eating."

Ben waited for the resident to come back down. The minutes dragged, one into the next, and still his resident hadn't returned. Ben suddenly realized the rapid response had to be for one of the less stable patients on their floor—which could mean it was for Mr. Moore.

Ben bolted upstairs. He reached Mr. Moore's room, where a cluster of doctors was gathered around the bed. Ben looked to the monitor: no heartbeat.

He stood staring at the body in front of him, which one hour earlier had been breathing and chuckling with his child on the phone. Mr. Moore had crashed abruptly; in this case, there was nothing else the team could have done.

Ben realized now that all the procedural steps were his to take as the primary provider. He had to record the time of death and sign his name with that new formal suffix, MD. He had to fill out a discharge summary explaining the care the hospital team had offered and call the number for LiveOnNY, the organization that would assess whether Mr. Moore could donate his organs.

The hardest part was picking up the phone in the resident workroom to call Mr. Moore's wife. It was Ben's first time following the script and it didn't feel sufficient for the weight of the words. If Mr. Moore's family were in the room, if he could look them in the eyes, maybe it wouldn't feel so hollow.

"I'm very sorry, but I have some bad news." Ben tried to steady his voice. It wasn't his turn to feel grief, not until he gave the family space to share theirs.

Mr. Moore's wife started sobbing on the other end of the line. Ben's head swam with the knowledge that this was his job now. To call up someone and tell them he'd done everything possible, but he couldn't save her husband's life, and to listen while she cried because she hadn't been in the room when he died. This wasn't a dress rehearsal in scrubs. No one higher up was checking over his shoulder to make sure he said the right things.

"We did everything we possibly could," Ben said. It helped that he knew this was true. He'd run through every lab. He knew his team had asked the right questions.

Mrs. Moore was breathing heavy on the other end of the line. She paused, fighting for her next words, and in the ensuing silence Ben felt the sting of her grief like it was iodine, sharp and unbearable. "Should we call a funeral parlor?" she asked.

Funeral homes across the city were backlogged. They hadn't been prepared for the surge of Covid deaths. Ben also knew that large in-person funerals were off-limits, so Mr. Moore's family would have to mourn virtually. Ben had seen articles about Zoom funerals, mourners sending their condolences through the cold airways of a chat box. It was another cruelty of the Covid era.

Ben read out next steps for Mrs. Moore and gave her phone numbers to call. He could sense her anxiety. If only he could take the logistical weight off her shoulders while she processed the gaping hole that had just materialized in her life. In school he had studied the technical aspects of death, but now he had to face them up close. People were handed a list of to-dos just when they needed most to hang up the phone, squeeze someone nearby, and give way to their grief.

Ben took home some hospital leftovers for dinner that night, and soon afterward he collapsed in bed. He kept thinking about the letters MD, which he had been working to earn for so long. He hadn't imagined that their first use would be for a death certificate.

Eleven

IRIS, May 2020

No one was under any illusions about the survival prospects of a patient hooked up to a ventilator. Some Montefiore residents said that in the first weeks of the outbreak, half of their ventilated patients died. That was the knowledge they had to hold, an impossible weight, when introducing themselves to their patients struggling to breathe.

Iris was assigned a patient named Mr. Johnson with a severe case of Covid-19. When she went to greet him, she saw a large man, African American, who had been given a tracheostomy, a hole cut through the neck into his windpipe to allow for easier connection to a ventilator. It looked like a plastic collar encircling the neck, with gauze underneath. It could be somewhat safer than normal intubation because it prevented scarring and ensured he couldn't instinctively knock the uncomfortable tube out of his throat, accidentally cutting off his own oxygen supply. But it was often used for people who needed to be on a ventilator for a longer period of time, and could suggest a more complicated process for recovery. Mr. Johnson was also on heavy sedatives and fentanyl.

"Hello!" Iris called out, making sure to raise her voice above the noise of the floor. "I'm Iris, I'll be taking care of you."

She was greeted by silence and hollow eyes. The machines around her patient beeped and whirred like a conversation in which he couldn't partake.

Iris was somewhat surprised to read that he was marked full code. His last doctor had spoken with his sister, who was adamant that if his lungs or heart failed, she wanted any intervention that could possibly save his life. They had their marching orders: do everything. This didn't entirely make sense to them, given his medical state, and they had been clear about his slim chance of a meaningful recovery. But it wasn't their place to overrule his family.

Iris came by again in the afternoon, when there was a lull in her shift. "How are you?" she asked him. "I'll be here checking up on you." No response.

He couldn't make eye contact when she spoke, which made Iris feel all the more ridiculous for her bubbly greetings. But these seemed necessary, too. "I'm always around," she told him. "If there's anything you need."

She noted his left hand dangling downward in an awkward position, so she leaned forward and lifted it gently. "I'm going to move your arm back," she called out. "Just to make sure you're comfortable."

Iris had cared for other ventilated patients who were more responsive. One woman, when asked to lift her hand up, mouthed to Iris: "I'm trying." Then Iris asked: "Are you in pain?" The woman shook her head no. "Is anything hurting or making you uncomfortable?" Again, no.

At the beginning of medical school, when Iris was dizzy with all the abbreviations she was meant to memorize, she had resolved that she would master the clinical vocabulary but not its tone. She was unsettled, specifically, by the patronizing way in which some physicians addressed their patients. She remembered her own diagnosis with depression, and knew she wanted to be the type of doctor who made her patients feel like partners in the medical process. Subjects, not objects.

This was tougher with some of her Covid patients, like Mr. Johnson. They couldn't speak and wouldn't even show a change in facial expression when poked and prodded by their nurses and doctors. It was difficult to know whether they comprehended anything the medical team said. But in her third year, Iris had started reading the medical literature on patients capable of hearing even in altered states of consciousness. She became consumed by the idea of patients slightly cognizant of their surroundings who weren't fully told what their doctors were doing to them. The loss of autonomy seemed so painful. Iris decided that regardless of whether her patients could engage her, she would always provide them with information on their conditions and care. Now, during Covid, that seemed even more necessary.

Iris struggled to decide what she should tell Mr. Johnson's family by phone. She didn't want to give them any measure of false hope, and his prognosis was poor.

"Tell him I'm praying for him," his sister said on the phone. Iris promised she would.

In what was both good and bad news for Mr. Johnson, Iris's team soon decided it was time to wean him off fentanyl, as well as sedatives. Opioids suppressed the respiratory drive—the body's compulsory fight for oxygen—and they wanted to see if he could start to breathe on his own. This was important, because his family wanted everything done that might help him recover, which meant getting him off supplementary oxygen at some point. Iris and her team began decreasing his dosage, with the goal of tapering it off completely.

His first day off the pain medication was a Thursday morning. He was Iris's first patient on pre-rounds, and she had to wake him to his first nonsedated, nonmedicated morning. She touched his shoulder gently.

"Mr. Johnson!" she called out.

No response. Still no eye contact.

"Your family has been praying for you," she told him. "They said to tell you how much they love you."

For the first time since he came into her care, Iris saw him somewhat register her presence. Then, to her disbelief, the man's eyes began to well up. A single tear spilled over onto his cheek. For a moment she was overcome with a desire to look away, to fix her eyes on anything but his face. He still couldn't hold her gaze, but his brow was furrowing, like he was beginning to come back into his own body.

Iris grabbed his hand. "Oh, I know," she said. "I know it's hard."

Oh God, she thought. Why had she been wishing for him to wake up? She began to realize how merciful that state of nonresponsiveness was, when he couldn't feel or understand his body's condition. But there was no going back to that now.

All she had wanted was for Mr. Johnson to gain some awareness of his surroundings. But in some ways, it had been easier to care for him when he was just a body in the bed, still and medicated. When all she had to do was follow his family's wishes, secure in the knowledge that he didn't feel the intensity and scale of his situation. But maybe this was part of her job, bearing witness. There was little she could do for him in his anguish.

After a few minutes, Iris had to leave to round on other patients. Outside, in the hallway, she pulled out her phone and texted her partner Benjamin: "If I'm ever vented, please just let me go."

Iris could manage pain and grief, those she'd experienced. But the loss of control that she saw in Mr. Johnson wasn't something she could handle, this lack of ability to understand or communicate.

Later, back at home, she described that teary expression to Benjamin. She told him about watching Mr. Johnson's face and wondering what he could make sense of and what he couldn't.

She and Benjamin were both young and healthy, so she knew rationally that if one of them was on a ventilator, they had a fair chance of survival. But she made Benjamin promise that if something ever happened to her, he wouldn't let her stay hooked up to a machine for too long. She didn't feel like she could handle a life without autonomy, without her own voice.

Their morning table rounds went by the SOAP method. That meant for every patient you discussed the Subjective (what happened overnight and in the morning); the Objective (their vital signs and labs); Assessment (summarizing their symptoms, conditions, and needs); and Plan (what should come next).

When Iris gave the update on Mr. Johnson, her team was dubious. She was certain that he was slowly becoming more responsive. It had been five days since he was taken off the pain medication. Sometimes when Iris rounded on him he was stiff, vacant. But other times, she swore she could see him respond to her voice. Once she asked him to wiggle his toes.

"Can you move them?"

And there one of them went, just less than an inch. Iris cheered for him. "Great job!" she told him.

Just last week he hadn't even seemed to hear her voice. It was hard for Iris to believe that he could now comprehend her instructions, so she tried just once more. "Can you move your toes another inch?" There they went again.

"He responds to you?" the attending physician asked her. Iris could hear in his voice that he didn't fully buy this. He thought she was imagining things. This attending had a reputation both for his dry sense of humor and for being one of the strictest teachers in the hospital. Most of the residents were terrified of him.

Iris told the team how Mr. Johnson's eyes had welled up when she first took him off fentanyl. She told them how his toes wiggled. She was conscious not to let her voice shake, not to get worked up.

You had to speak the physician's language of flat logic. This felt especially important for Iris as someone brand-new to the team and the wards.

None of the residents had seen him acknowledge their voices. "Are you sure it's not wishful thinking, Iris?" her attending physician asked, his tone good-natured. "It happens to everyone."

Iris had to clamp her lips shut so she wouldn't protest. She felt a flash of doubt: Could she be imagining things? She'd been so confident that the movement of his toe had been in response to her voice, but maybe the weeks on the ward were wearing on her. Maybe the team thought that she was starting to crack from the pressure.

A few days later, Iris was at her desk, reading through charts. Her attending appeared wearing a peculiar smile. "You were right, Iris," he said.

"About what?"

"I saw Mr. Johnson respond to me."

Iris almost wanted to do a victory lap around the room. Instead she just offered a slight nod and thanked the doctor for the update. But she felt a wave of vindication. In the last few weeks on the wards, she had felt the pangs of inexperience. She'd wondered whether she was being judged, whether anyone noticed her voice shake when she reported a patient's vital signs to her team. But she also knew instinctively that she brought a different kind of skill to her job, one earned less through experience and more through attention: the capacity to listen, even when the patients couldn't speak. Anyway, so much of a doctor's work was about tolerance for risk and uncertainty, and Iris was pleased to know that the trust she put in herself wasn't misplaced.

The lifesaving nature of modern medicine lies partly in its reliance on order. Doctors know where they stand in the hierarchy: attending physician, senior resident, resident, then lowly intern.

As on a ship, when seniority dictates division of labor, all stays afloat. There are clear channels of accountability, and people tasked with overseeing the decisions of those lower on the chain. But crises tend to scramble the logic of routine.

In the case of Covid, as it spread across the country, what that meant was a shift away from typically rigid lines of command. Take Yale New Haven Hospital, which assembled a list of four hundred physician volunteers to come help in the wards. The dean of the medical school was one of the first to enlist. Physicians who normally worked in operating rooms weren't able to do elective procedures, so some switched to helping with Covid patients. Dr. Lina Miyakawa, a pulmonary critical care physician in New York, put it aptly: "Surgeons were drawing labs or cleaning up somebody after they urinated. That's definitely not their job, but everyone had to pitch in."

Nowhere was the collaboration on sharper display than at Bellevue Hospital, which saw some fourteen hundred Covid-19 patients during the city's first wave. There were twenty-five codes daily, compared to the normal average of one. Bellevue's ICU typically houses twelve patients at a time; in spring 2020, that number jumped to nearly a hundred. Twice-daily briefings allowed physicians to share information about protective equipment and experimental treatments like hydroxychloroquine and remdesivir. And, in a show of communal spirit, doctors agreed to shuffle across departments. Some moved from outpatient clinics to internal medicine, others from elective surgery units to the emergency department. For all the trauma that the early graduates witnessed, they bore witness to an unusual degree of cooperation among their colleagues.

The trainees also had the benefit of gaining early expertise in a disease that was novel to all of their colleagues, not just the rookies on the floor. Especially in those early weeks, each one of them faced the same knowledge gaps: How did the virus spread, what were the most common symptoms, what worked best in treating it? And new information emerged each day—that the virus transmitted more by aerosolized particles than by fomites (surfaces), that hydroxychloroquine might cause just as much harm as good. The residents deepened their expertise alongside their supervisors. That meant, as Iris experienced, that a newly graduated medical trainee might have nearly the same understanding of a Covid patient's trajectory as her attending physician.

In some instances, that novelty could soften the hubris a more senior physician might bring to their work. "Medicine tends to be hierarchical and ego-driven," explained Dr. Lynn Fiellin, an internal medicine doctor at Yale New Haven. "So much of that went out the window." She predicts some of those shifts will stick even as the pandemic era fades. "We created ways of doing things that will probably be sustained, because they're better than what we were doing before. There was a sense of common mission."

Loosening some of the rigidity in medical lines of command might actually benefit people's health. Medical error is a leading cause of patient death in the United States, and when residents or interns don't feel comfortable critiquing their superiors, those errors are less likely to be caught. It's also imperative as the field grows more diverse. New physicians are trained in patient-centered care and cross-cultural competency in ways that older doctors often weren't. A senior physician with three decades of experience might recognize the signs of appendicitis within seconds—pain in the belly button, rebound tenderness in the abdomen—yet lack some of the sensitivities around medical communication that their younger colleagues have cultivated. Older white physicians might be particularly ignorant of the way racial bias affects the full axis of medical care.

The problem is that while young doctors are instructed to listen and learn from their superiors, the teaching doesn't always flow the other way around. Medical hierarchies don't tend to leave much space for a resident to tell an attending physician "You're wrong." For years, one of the difficulties of changing medical culture has been the conformity bred by hospital power structures. Medical students might learn some new, innovative technique for sensitive caregiving in medical school, but when they arrive on the hospital floors they are nervous about challenging sources of authority. Perhaps they chuckle along with an older physician's off-color jokes rather than questioning the biases they encounter. They watch the older residents, too, who don't dare undermine the most senior physicians. They unlearn their lessons on sensitivity.

During the Covid weeks, those hierarchies began to bend, ever so slightly. Teams were re-arranged, and doctors pitched in where they were needed. "We were all going through this mass trauma collectively," says Dr. Rana Awdish, the Detroit critical care physician. "It was an environment that allowed us to access our emotions and be more transparent than we normally are. When I think about the things from this time that I want to carry forward, it's that leveling."

When norms shift and vanity dries up, there can be unusual opportunities for communication. Medical providers are all co-operating to navigate unfamiliar territory. Ego gets displaced by a need for efficacy and efficiency. And veterans realize that they can learn from their recruits.

ELANA, May 2020

Like an office zealot who commandeers the copy machine, the coronavirus hijacks a healthy cell and begins replicating. As the infection travels the body's airways, the immune system fights back, mustering a response that causes its own collateral damage alongside the wreckage from the virus. Fluid from the blood vessels leak and the air sacs in the lungs fill, blocking them from taking in oxygen. The lungs swell and stiffen. On an X-ray, they sometimes appear overtaken by hazy areas called opacities that look like the frosted glass of a shower door, blurring but not entirely concealing what's underneath.

These were the lungs, filled with debris, of one of Elana's patients who had suffered Covid pneumonia. He was being weaned off oxygen and was finally getting ready for discharge. But during a meeting known as interdisciplinary rounds, when staff members from across different departments met to discuss their patients, the social worker told Elana's team that his insurance provider refused to cover the costs of rehab after his ICU stay because he wasn't from New York State. Elana knew he would have to relearn basic motor skills after critical care—how to walk on his own, breathe on his own, dress himself. He was shaky from sedative medication and loss of oxygen to the brain. The physical therapy team had evaluated him and documented his needs, thinking he'd be sent to a rehab facility.

The team didn't want to get into all these demoralizing details with the patient, especially while he focused on recovering. Elana listened to her attending physician feud with the insurance company for hours. "Do you understand how long he was on a breathing tube?" The company wouldn't budge. Elana's attending physician decided they would keep him at the hospital a few more days for inpatient rehab, which carried its own risk. Every extra day left the patient vulnerable to some other infection floating through the hospital halls. It was a trade-off.

Watching her attending, Elana realized that this was the less glamorous but no less important side of their job—endless paperwork and negotiations on finances. What Elana found most affecting, in her first few weeks, was the way her colleagues could absorb themselves in these logistical considerations and remain prepared to shift into poignant conversations on patients' emotional needs at any time. They had to speak the language of Medicare and mortality all at once.

Elana had always had this sense of wonder at the work of medical providers. She had a poster taped to her bedroom wall that depicted a debate between a physicist, a chemist, and a biologist about whose field was most important. The biologist pulled out a trump card: a chart showing deaths from infectious disease throughout history, which plunged to near zero in the modern era. "The giants of our field have slain the horsemen of the apocalypse," the biologist declared. This was their work, just short of miracle making.

Elana had a middle-aged patient, an African American woman named Ms. Spencer, admitted with Covid-19. She didn't need intubation, but Elana could tell she was petrified that her breathing would worsen. She tried to force a sunny tone in her exchanges with Elana, though her downcast eyes often betrayed her real emotional state. Sometimes when Elana came by, she was on the phone with her partner, who couldn't visit because of the hospital's Covid rules.

One evening, after Elana had finished checking in on her other patients, she stopped by the room, and Ms. Spencer asked shyly if she could request a favor.

"What can I do for you?"

"Can you make sure I don't die?"

Elana froze. Usually she prided herself on knowing how to respond in delicate situations, but this felt different. She was acutely aware that there were no other residents and attending physicians around. This patient was looking to her for some properly calibrated response, and Elana was feeling all the weight of her worries.

It just so happened that when Elana was driving to work that morning, she had turned on the car radio and heard a tribute to the frontline medical workers. The segment featured a nurse telling the story of a patient who had asked if her life could be saved. The nurse had responded, "I can promise you that we will do everything that we possibly can." Maybe these were the right words to parrot, Elana thought, feeling thankful for the blueprint.

"I'm scared this is going to kill me," Ms. Spencer continued.

It was jarring to hear someone, especially a patient who'd been so outwardly optimistic, say out loud what everyone in the hospital had tangled with in private. Suddenly all their fears took on the solidity of sentences and sound. For all of Elana's angst in the weeks leading up to her start date, once she had started on the floors, there hadn't been much time for airing out apprehensions. There was too much work to be done. Now Elana could feel her mouth start to form the syllables of a feel-good response. "Of course I'll make sure." Or "You won't die, I promise." These sentences were easy enough to get out. But they weren't honest. There was so much they didn't know about Covid. Elana thought back to the bad-news simulation with Iris and the medical actor, reminding herself not to gloss her words in faux comfort. Just truths.

"This is a very dangerous disease, you're not wrong about that," Elana said. "But we're going to keep a close eye on you here. And if anything goes wrong, I have high hopes we'll be able to catch it."

Ms. Spencer's milky eyes were still searching. Elana wished that she could know every detail there was to know about this patient's life. As with the cadaver who had those pink-polished nails, having a narrative helped. The care was built on stories, not just clunky medical terms.

"We're going to do the best we can for you," Elana said. "I can promise to you that we will do everything that we possibly can." It occurred to her that Ms. Spencer might want to take the opportunity to talk about her goals of care, and go over her plan in case the disease worsened. "Do you want to talk about what we would do for you in case something really bad happens?"

Ms. Spencer nodded. Elana asked her whether she would be open to being intubated or resuscitated in case her lungs or heart failed. The two discussed what it would mean to receive chest compressions or be put on a ventilator, the risks involved with each, and Ms. Spencer said she wanted to be full code. That seemed like a reasonable choice: Ms. Spencer was middle-aged and in fairly good health. Still, as Elana excused herself and left the room, she knew that back home she'd be worrying about the woman's oxygen levels. She had read somewhere that one of the biggest predictors of death was when the patient thought they might die.

It was late. Elana was hungry, and Akiva probably had dinner ready. She stripped off her scrubs in the changing room and wiped down her ID badge with disinfectant. As she was walking toward her car, she got a text from her dad.

"How are you?"

If they were together, she could tell him the details of her day, relaying the story of the woman who asked not to die. But that was too long and sticky to translate on text. So she sent back just a single word: "Good."

"Do you like your job?" he wrote.

"Yes," she typed.

His response came immediately, and it made her smile: "Well that's a gift."

At some point, she'd get to see her dad in person again. After she finished her Covid care weeks and before she started residency, she planned to make another trip to visit her family on Long Island. Her dad had also mentioned maybe they could hike the Appalachian Trail. But that felt far away. For now, she just wanted him to stay safe.

A month earlier, before the outbreak shut down places of worship, he had planned to go to a celebration at their synagogue for the Purim holiday, and Elana had called, begging him to stay home. "This Covid thing is worse than people think it is," she told him. "You could get really sick." She remembered almost hearing his signature good-natured smile on the other end of the phone as he told her, "You might be making a bigger deal of this than you need to be."

Since then, their roles had begun to reverse. His sense of alarm kept rising steadily, while Elana's had begun to mellow now that she had reached the wards and started settling into work. It felt to her like the coronavirus was the monstrous villain of a low-budget horror movie. Her family members were watching the shadow it cast on the wall lengthen as it drew nearer—What was it really? What if they had to face it?—while she and her fellow hospital workers had already stared it directly in the face. They might still be afraid, but at least the mystery had disappeared.

Twelve

JAY, April 2020

Some days faded together, a haze of charts and codes and high-pitched calls. Then there were moments so vivid they might have been taped and replayed on a loop in Jay's head. The week she met Manny was one of those.

It was a slow morning. Jay was hunched over her computer in the workroom. When she heard shouting in the hallway, she sprang up and ran outside, looking for its source. There was a Hispanic patient with Down syndrome, wearing the hospital's typical striped blue gown, who was ripping binders off the nurses' station and throwing them on the floor.

"Stop!" one of the nurses yelled, frantically grabbing the binders off the ground and setting them back in their place. "What are you doing? Stop!"

Her pleas went unacknowledged. The *thud-thud-thud* continued as the patient dropped reams of paper. He was wailing, his cries mounting to a crescendo that flooded the hall. A second nurse appeared and shushed the first. "He does this when he's frustrated," she said. "Don't worry."

She put out her hand. The patient eyed her outstretched arm

warily, then grabbed it. Slowly, he followed her back toward his room.

That was Manny, whom the doctors called by his first name, unlike many of the other patients. He was slight, just five foot two, with dark hair, thick brows that curved downward, and a five o'clock shadow on his chin. He often got emotional, and sometimes disruptive, but usually all he wanted was touch, someone he trusted to grab his hand. Quickly, willingly, Jay would assume that role. (She would later learn all its technicalities. Like when Manny got agitated, some of the doctors gave him psychotropic medications, which were given pro re nata, "as needed." But others held back, because each time he was given these drugs they were documented in his records and could decrease the chance that a group home for the developmentally disabled would take him in. Often he didn't need medication so much as some soothing words.)

The patient was assigned to Jay's care that week, and she heard his story in hushed conversations with the nurses on the floor. Manny was thirty-eight. He came to the hospital when his father, who was his only living immediate family member, was admitted with Covid-19, riding over with him in the ambulance. Manny's dad soon coded with his son at his side and was put on a high-flow nasal cannula in the ICU.

The hospital bent its rules on visitors and allowed Manny to stay; there was no one else who could watch him. Manny's mother had passed away a few years earlier. He and his dad had been living in a city housing project with little connection to the outside world. Now Manny spent his days seated by his dad's hospital bed. The nurses searched his face, wondering how much of the situation he could comprehend. His dad communicated with him in Spanish, so they weren't sure whether Manny could pick up on much of their English.

"Is there anyone we can call to care for Manny if something happens to you?" the social worker asked Manny's father one afternoon.

Manny's dad shook his head. "I've always planned on outliving him."

One night, around 4:00 a.m., his dad's heart stopped. The doctors rushed in and tried resuscitation with Manny lying there in the room. They did everything they could to no avail. Manny's dad was declared dead. The overnight staff wrapped up his body and prepared it for delivery to the morgue. Shortly after Manny was swabbed, tested positive for coronavirus, and admitted to the hospital as a patient. Manny was put on one-to-one care, meaning he always had a staff member with him.

Down syndrome, a condition caused by an extra copy of chromosome 21, can cause a variety of disabilities, but Manny's were severe. Some of the hospital's doctors and nurses were frightened by his tantrums. The psych resident who worked with him reminded the others to show him empathy and not to get upset when he threw things on the ground.

Gradually Manny became a favorite among the hospital staff. He was affectionate—when a nurse or doctor spoke to him patiently, he grabbed their hands, his warm eyes opening like a welcome mat. "Be patient," the other providers warned Jay when he was assigned to her. "Don't take it personally when he acts out." But all Jay could think about was the relief on Manny's face when the nurse in the hallway took him by the hand. Jay wasn't too nervous about caring for him.

She stopped by Manny's room to introduce herself. He was propped up in his small bed, which was just slightly wider than him. His room was sparse: a wall painted a washed-out robin's-egg blue, a single brown plastic chair. There were a few toys scattered around—a plush football, a stuffed duck—donated by hospital staff. One of the doctors had bought him a mini basketball hoop, and his social worker Alicia had brought him *Dora the Explorer* picture books.

"Hola, Manny," Jay said. "Soy su doctora nueva." He swiveled his head and eyed her cautiously. She moved toward his bed and put a hand on his shoulder. To her surprise, he didn't shake it off.

Later, just before her shift ended, Jay stopped by Manny's bed to say goodbye for the evening. He was sitting up and crying quietly, clutching an ID badge on a string around his neck, which he kept looking at while his eyes welled up. Jay stepped toward him, and he angled the badge so Jay could see the photo fixed to the back of it.

"Papi," Manny said, pointing at it. It was a snapshot of him and his dad. The two were wearing matching shirts, white with blue stripes like Yankees uniforms.

That night, on the way home, Jay thought about Manny lying in his hospital bed alone, fighting to fall asleep and fighting off thoughts of his dad. Then she called her mom.

Manny loved to dance. In the hallway, when he went for a stroll, the nurses played Spanish pop artists like Selena or Enrique Iglesias on their iPhones, and he drummed his hands like he was hitting a snare. From the corners of his mask, Jay could see his lips turning up into a grin as he grooved.

Manny's medical records marked him "nonverbal." In their first days together, the only words Jay heard him say was "Papi," when he looked at pictures of his dad. Or "No," usually responding to nothing in particular. They couldn't exchange sentences but they found other ways to communicate. He pointed at objects around the room, and Jay offered a running commentary in Spanish.

"Si, ahi esta la ventana," she confirmed, following his finger toward the window. Because Manny had recovered from Covid and had antibodies, Jay didn't have to worry about holding his hand or letting his head rest on her shoulder as he napped.

Jay was beginning to understand that some patient-doctor relationships are sewn of a fabric thicker than words. It was a trust bred through time, but mostly through effort. Jay wanted them to be close. Manny could tell. When her shift was over and her tasks done for the day, she came to Manny's room and took him on walks around the floor.

The cap on resident work time—eighty hours a week—was lifted for the Covid months, so by the time Jay got to the floors her coworkers were exhausted, frayed by crisis and grueling shifts. She was still fresh, ready to give the sort of deep care that Manny needed.

Manny healed quickly from his case of Covid, which wasn't severe. But his release from the hospital wasn't straightforward. When he first turned up, he'd been in his father's care. Since his dad died, he had no one to speak for him. The possibility of going home to his former life had died with his father. One of the hospital's head social workers, Alicia, was working on developing a safe plan for his discharge. There were complex questions involved—where would he live, who would make his medical decisions?

Alicia, fair-skinned and dark-haired, was a fiercely intelligent woman who spoke in high-voltage sentences, her every command charged with urgency. She was petitioning for an emergency court hearing so Manny could be appointed a legal guardian, and she had also begun filing applications for him at New York group homes for people with developmental disabilities. This was a complicated process at any time, but even more difficult during the pandemic. The homes were wary of interviewing someone who had just spent a month at a Covid-heavy hospital. Most of them had a wait list of two or three years, though Alicia had learned that some had unexpected openings because of residents who had died from Covid.

One of the homes with an opening was in Corona, Queens. They arranged for Manny to come interview in person on a Wednesday afternoon. Before leaving the hospital, Jay and Alicia changed into clean scrubs and new masks, sanitized their hands, then called an Uber with one of the hospital's psychiatrists and a patient care assistant (PCA).

Manny was delighted to be outside. After five weeks in a hospital bed, New York streets can seem like the promised land. The windows of the redbrick buildings along First Avenue glinted in the afternoon sunlight. There was the bodega across the street, the liquor store with its neon-red-and-blue Wine & Liquor sign, the yellow awning of Discount City. There was the little silver cart that sold lukewarm coffee and bacon, egg, and cheese rolls. The block was tucked into the calm that came from quarantine, all its usual late lunchers and cell phone barkers shut in at home.

Manny didn't seem to know where they were headed, but he was happy on the ride over. He gazed out his cab window at passersby as Jay made friendly conversation with Alicia, the psychiatrist, and the PCA.

They arrived at a residential-looking brick building with a wrought-iron gate and a wheelchair ramp outside. They were greeted at the entrance and deposited in a homey living room with a faded rug, a welcome departure from the cold linoleum of the hospital floors. As they were seated on a sofa, Manny seemed to realize that something formal was happening, and he balked. His eyes turned accusatory. Jay stroked his arm to keep him calm as he burrowed into the couch cushion.

After some time, Manny stood and began to wander around the home, with Jay and the others following behind him. Nine of the home's residents were scattered around the dining room talking, and Manny sat down to watch them with nervous curiosity. Jay had forgotten what it was like to be in a room thrumming with social activity, one with no terse voices or respiratory machines.

One of the group home residents spotted Manny and approached. He stuck out his hand to shake Manny's. Then the young man gestured toward a cluster of home residents nearby, as if to say, Come join. Manny's face broke open into a wide, genuine smile—the sort he usually had when he was grooving in the hallways.

Jay hadn't ever seen Manny with peers. She only knew him in the context of his hospital room. There was a lightness to him as he bounced around the group. She was struck by the way geography can define someone, how a certain space can bring parts of our identities into sharp focus and leave others blurred. Back at the hospital, Manny was a patient or a social work case. Here he was just another young person, looking happier than he had since Jay met him.

Manny left the group at one point to climb the stairwell and poke his head into one of the bedrooms on the second floor. Jay could see the expression of desire, almost need, fixing on his face as he took in his surroundings: this place was a home.

Two or so hours passed, and then Alicia gestured to her watch. It was time for them to leave. Manny's body went slack. His eyes fixed on the doctors, and he started shaking his head vehemently.

"Manny, we have to go back to the hospital now," Jay said. "I know it was great meeting these new friends. But it's time to go back now."

Manny backed away from the doctors. He continued to shake his head: *No, no, no.* In his hand, he gripped the necklace with his father's photo like a shield, like an amulet. "Papi," he cried.

The home's other residents were quickly whisked away. As the room emptied out, Jay felt the emptiness in Manny's life viscerally, the loss of his dad and his normal life.

Jay, Alicia, and the psychiatrist tried coaxing him toward the exit. So did the group home's staff, with whom he had bonded rapidly. One of the group home aides even offered to drive him back to Manhattan in her car. But none of their pleas worked. Manny did not want to leave. He kept ducking away from them, shaking, puncturing the air with shouts of "No!"

"We're going to have to call an ambulance," the psychiatrist said quietly.

They called 911, and an ambulance arrived in what was actually minutes but felt interminable. Alicia begged the medics not to sedate Manny as they gathered around him. Manny burst into tears as he was lifted onto a stretcher. Jay finally had to look away. She kept seeing flashes of his proud smile when the first group resident came over to shake his hand. "No, no, no," he kept shouting as he was strapped down.

"Thanks for your time today," the group home director said to Alicia and Jay, with a dose of distant pity. "We'll evaluate his case and get back to you next week."

Jay and Alicia exchanged a look, their eyes welling up. It seemed like they would be back to the drawing board on the plans for Manny's release.

Manny wouldn't meet Jay's gaze that evening at the hospital. She tried everything—cheery greetings, an apologetic smile, a quiet stay sitting near his bed—but he just stared stone-faced at the wall. She could sense his message: *Traitor.* After a few minutes, Jay left him to rest. The only currency that she and Manny traded in was trust, and it wasn't in inexhaustible supply.

Jay was transferred to a new unit the next week. She wanted to go where she was helpful, but she was disappointed to leave the patients who had come to occupy so much of her mental and emotional focus. Mostly Manny.

Manny was still interviewing at various group homes, though mainly on Zoom. Some of the video interviews were difficult. Manny didn't understand why he was being asked to interact with these pixelated faces on a computer. Sometimes he just stared. Sometimes he was stirred by disturbances on the other side of the screen.

"Gato!" he cried once, pointing at a cat that had wandered by the female interviewer's video camera. Jay and Alicia looked up in surprise. They hadn't realized that this word was part of his vocabulary. Jay found herself wishing the group home directors were there in person to see the charm in his gaze. In each conversation, Alicia emphasized all the activities Manny was capable of doing on his own: walking, going to the bathroom, getting dressed. She sent over iPhone videos of him dancing.

As she prepared to transfer teams, Jay tried to tell herself that Manny would move into the care of some other doctor who would be equally consumed with helping him. But she worried that she would feel his absence like a phantom limb. She wondered if his bouts of anger and tears might erode some other doctor's patience. Manny cried most often when he saw something that reminded him of his dad. Then he would shake his head in confusion and grip his ID badge while calling out "Papi!" But it was the unspoken rule of the hospital: patients were your everything until they weren't.

Before Jay transferred, she decided to clear time to go through Manny's medical chart and make sure she hadn't missed any notes. She pulled up his records on the workroom computer. If she was going to leave him, she wanted every detail of his history dissected.

Each patient admitted to the hospital had a medical record number (MRN). Some might be accidentally assigned a new MRN when admitted, even if they already had one. This could happen if they gave a new version of their name, or if their date of birth was entered incorrectly. Manny had been assigned a new MRN when he got to the hospital, so his medical history was blank.

Alicia had worked to gather any data point she could on Manny. She had interviewed his dad before he died, standing just outside the ICU room and talking through a Spanish interpreter as the dad ripped out his oxygen tube to ensure he was being heard. She had taken down Manny's Social Security number, allergies, anything she could think to ask. She'd also found his file in the state's Office for People with Development Disabilities, but the record had little information because his father had only brought him to one or two sessions with a case worker in 2012.

But Jay also knew that her facility used a different electronic records software than the other main hospitals in its system. The rest used Epic, a digital medical system run by a health care behemoth based out of Madison, Wisconsin. Jay's used a system called Prism. The hospital was meant to transition over to match the other facilities that spring, but the deadline had been pushed back because of coronavirus.

Jay began to wonder what would happen if she searched Manny's date of birth in Epic. She scrolled, slowly. And suddenly there it was: identical birthday, identical address, name rendered slightly differently but still there. She clicked and began to read.

The Epic file was a gold mine of information. For one thing, Manny was partially deaf. At one point he'd been fitted for hearing aids, which were presumably still in his dad's empty apartment. He also had a generalized anxiety disorder. He'd been prescribed selective serotonin reuptake inhibitors (SSRIs), which he hadn't had for all his weeks as a patient. Jay's mind scanned through the weeks of tantrums and tears, all chalked up to the stress of his hospital stay and the loss of his dad.

She sprinted down a flight of stairs to the workroom. It was a tiny space, just five by eight feet, with a single window and four computers squeezed in. Her resident and two co-interns were behind their computers, and they looked up as Jay flung the door open.

"Is everything okay?" her resident asked.

"It's Manny," Jay said, bending over to catch her breath. "He had another MRN."

"What?" The team members looked at Jay and then one another.

There was silence for a minute, so Jay continued. "He's profoundly deaf in one ear and almost deaf in the other." She watched this news register on her resident's face. "He's also been on an SSRI for over ten years."

They all froze. This was nearly impossible news to process.

For all the time and energy they'd poured into Manny's care, what they had been missing was data—fundamental conditions he suffered from, which had nothing to do with the reason for his current hospital stay but everything to do with its quality. Jay processed with new clarity what a precious resource information was. She'd been under the impression that in these weeks, time was everything, that there was a linear relationship between hours spent and efficacy of care. But the equation didn't balance without medical history—time could simply be misdirected. She'd been so focused on the language barrier, trying to give instructions in Spanish and wondering how much Manny could understand. Now she wondered how much he'd even heard.

Patients who spent weeks in the hospital liked to joke that they found their Covid families—the nurses they saw daily, the doctors whom they nicknamed. But the providers needed these families just as much. Far from her own parents, limited to brief evening interactions with her med school roommates, Jay found her whole world was becoming the hospital floor.

Her days were busy with Covid patients. Their symptoms were often similar—shortness of breath, sometimes blood clots—but their hospital stays never unfurled the same way. There was a mid forties woman whose husband explained as she was admitted that even when her blood oxygen levels plummeted, she'd been too nervous to go to the emergency room. She ended up waiting so long that she developed hypoxia and crashed, prompting her husband to jump in and do CPR. There was a woman struggling for breath who had been turned away at three emergency rooms, each time told that her symptoms weren't bad enough for her to be admitted. By the time she got to Jay's hospital, she was desaturating (low on oxygen) and had to be intubated within two hours. There was a man with prostate cancer who had been in and out of the hospital so many times that he and his wife were fluent in medicalese. They asked Jay if she was the intern or senior resident, well versed in the staff hierarchies that most patients didn't know to track. These patients were called the "good historians"; they could rattle off their medical conditions and drug prescriptions as though listing the members of their own family.

Jay was also getting acquainted with the staff. One morning, a surly cardiology fellow she worked with declared, "I don't want to know about your patients." Jay stared at him, taken aback. Doubling down, he added, "I only want to hear from you if there's a procedure I can do." On the other end of the spectrum, there was the affable resident who never seemed to tire of Jay's questions. What's the right dosage of blood thinners for a Covid patient who's clotting? And with that dose, what side effects should you look out for? (Covid patients were known to clot more easily, but because the disease was so new, the protocols for treating it weren't easy to determine.)

But she felt a certain tenderness toward the orbit around Manny's room. Those were the staff members who worked overtime, who placed the extra phone calls to make sure Manny was safe. They brought him new clothes and books. They found out his favorite foods and songs. One of the PCAs had even started to research the possibility of adopting him. It reminded Jay why she'd wanted to go into the field in the first place, for the patients who weren't just patients and the doctors whose work didn't fall neatly within clinical lines.

Even after her transfer to a new unit, Jay tried to visit Manny for at least twenty minutes each day, usually on her lunch break. In retrospect, she realized that those minutes were just as much for her own sanity, a reminder of the human side of the hospital. Manny had quickly gotten over his frustration with her from their visit to the group home. His face brightened whenever she came in. He always instantly, instinctively reached for her hand. He liked when she stroked his back, too. When she tried to leave and return to work, he turned a set of eyes on her that made it hard to go.

"Oh my God," Jay told Alicia one afternoon. "When he actually goes, I'm gonna miss him so much!"

Sometimes Manny had new acquisitions to show off—a blue collared shirt donated by one of the nurses, a stuffed penguin almost as long as his torso. He seemed especially delighted by this new guest, wrapping his arms around its flippers and patting its plush orange beak. Alicia had been able to arrange for a lawyer to get access to Manny's old home, and he'd returned to the hospital with a cart full of stuffed animals, clothes, and a CD player for music. During this visit—in which it became evident that Manny's dad had been a hoarder—the lawyer got trapped in the apartment and finally exited by climbing onto the fire escape.

After they'd found the details on Manny's medical history, her team was able to coordinate with a friendly resident at the nearby Ear and Eye Hospital to get Manny's earwax cleared so he could be fitted for a new pair of hearing aids. That meant Jay didn't have to raise her voice anymore when she came to visit. They also got him on anxiety medication, and his tantrums started to abate.

One afternoon Jay was sitting next to Manny's bed when a woman came in, dressed in street clothes instead of scrubs, a denim jacket over a bright pink tee. Her hair was tied high in a messy ponytail. Jay wondered if this might be a distant family friend, someone who had heard about his situation. Her booming greeting told Jay that she knew Manny had previously struggled with his hearing.

She turned to Jay and introduced herself. She was one of the nurses who had cared for his dad, and for Manny during the month he spent at the ICU bedside. It was her day off, and she wanted to see how he was doing. She circled around to the other side of the bed and grabbed his hand. He looked up and registered the familiar face. Manny didn't say anything, but his eyes flooded with new warmth.

During Jay's next visit, a middle-aged man, maybe five-foot-six, balding and tattooed, came by Manny's room. "Papi!" Manny shouted when he appeared at the door.

Jay froze for a moment. She knew Manny's father was dead. Could this be another relative, an uncle? It wasn't entirely out of the realm of possibility—caring for Manny was like opening a set of Russian dolls, each step forward exposing new layers of complication.

"Hey, papi," the man called back.

Manny's hand floated to the photo fixed onto his badge with his dad's face.

"I'm Paul," the man said, turning to Jay and nodding his head slightly. They couldn't shake hands because of Covid. "I'm one of the PCAs. He thinks I look like his dad," he added quietly. Jay could imagine being disconcerted by the mix-up, but Paul didn't seem to mind.

That afternoon, one of the patients across the way from Manny got agitated. His nurse was trying to finger-stick him to monitor glucose levels, and the patient began shouting racial slurs. As a second nurse ran over to calm him down, the man took off his pants and threw them on the floor.

Manny watched from afar. Jay positioned herself in front of him, blocking his view. She hoped that he couldn't make out what the man was saying, which was mostly a stream of expletives. She tried to distract him by pointing toward the window.

When Manny came to the hospital, the branches on the trees outside had been naked and frail; now they were beginning to burst into green, tipping their faces toward the spring sun. The city shimmered with a warm almost-summer glow, shrugging off the cold, cloudy days of March. Manny watched it all from his ninth-story window. It looked out mostly over another redbrick building, but it could offer a glimpse of the park if you tilted your head.

Thirteen

GABRIELA, April 2020

Gabriela switched to night shifts in late April—the early graduates were moving between shifts and teams, as they would during their internships. Her new work hours were from 7:00 in the evening to 7:00 the next morning. That was light, by medical standards; during residency she might work 5:30 p.m. to 7:00 a.m. But the twelve hours still felt physiologically hellish.

She fueled up on coffee before her first shift and biked to work in the fading evening light. Later in the night, some of her co-interns would nap on stretchers or chairs in the infusion room that was serving as their workspace, but Gabriela was too alert to sleep.

By day two Gabriela felt sleep-deprived and dehydrated. It was hard to remember to drink water, especially with all the layers of protective equipment that had to be removed. It became a recurring half joke among the interns that any queasiness or fatigue might be a symptom of Covid, rather than a sign that they'd spent too many sleepless hours on their feet.

Switching to night shifts meant swinging over to the opposite end of hospital workflow. Unlike doctors on the day shift, who were assigned their own patients and drew up plans for their treatment, the night shift kept their eyes on everybody and made sure that none of those plans went awry. They tracked vital signs and passed on relevant information to doctors clocking in the next morning. The early graduates worked with their senior residents, admitting new people who came in and checking in on the sickest patients throughout the night.

A local restaurant had donated sushi for the frontline workers. Gabriela went to pick up a roll and bumped into an old medical school friend who was also on shift that night, working a different floor. They hadn't seen each other since before they started in the Covid wards, and it felt almost like a confrontation with the ghost of her normal life. "You're here, too?" It was the closest thing to a social interaction that she'd been able to have in days.

In the morning Gabriela left the hospital and biked home to Little Italy. It was Sunday. In some other world, cycling through the streets just after dawn on a misty spring weekend might feel cinematic. But she was too tired to take in the hazy light, gripping her handlebars so she wouldn't topple to the ground. She always biked home in her scrubs, and she liked the looks of grateful approval she got from passersby; a lady on the street once yelled, "Thank you for all you are doing!"

Gabriela got home and locked up her bike near the neighborhood bar, which in ordinary times would have been recently vacated by its night crowd, miniskirted New Yorkers vaping while they waited for their Ubers. Up in the apartment, Gabriela made her way to the shower, as Jorge woke up and asked about her shift.

Weekends they liked going out for noodles at Xi'an Famous Foods, for tea at Cha Cha Matcha, for thick Detroit-style slices at Prince Street Pizza, or for sandwiches at Parisi Bakery, where all the workers knew their names and orders. They liked playing board games and watching the Nets. Now Gabriela was fighting to stay awake during their few hours together and punctuating her sentences with yawns. But one night, the week that she had once planned to be on a postgraduation trip to Madrid with a classmate, Jorge deemed it a celebratory Spanish-style evening and made tapas while Gabriela changed into a red dress.

Those early weeks of Covid were like a shrinkage. Lives usually filled with brunches and beers and nights out dancing retracted into tiny apartments. The streets and subways deflated, and the only things left were screens, and phone calls to family members all locked in their own deflated lives too. To Gabriela, being with family in person was like a real meal; talking to them over her iPhone was like processed food from a box, just filling the space.

Gabriela's mom was still battling for her small business loan. Over the phone, she explained that the federal Paycheck Protection Program had run out of money after less than two weeks of operation. Gabriela read news stories about the massive companies that had weaseled their way to federal aid. The money covered any company with fewer than five hundred workers in any one of its locations. Shake Shack made headlines after returning its $10 million loan.

Gabriela listened as her mom boomeranged between gloom and dogged optimism. "When we reopen, we'll have to be working seven days a week," she told Gabriela. It seemed to alleviate some of her current stress to plan, in exacting detail, the logistics of the return to work. "We'll have half the girls come in the morning and half in the afternoon, and in between we'll do a full clean. Sterilize everything."

It felt strange to know that all their lives were moving forward even as the world came apart. Jorge's brother and sister-in-law had just had a baby. Jorge and Gabriela met their new nephew over FaceTime one afternoon. Ever the pediatrician, Gabriela felt physically pained to see his body on screen without being able to squeeze his toes. It was even tougher knowing they were only twelve miles away across the Hudson.

She was glad that this infant had no awareness of the uneasy world that had just welcomed him. He certainly had no awareness of Gabriela and Jorge, faces appearing on a phone to coo and ahh. Jorge watched Gabriela watch the baby, and Gabriela felt this shock of tenderness like an ache in her bones.

In the hospital, it was called sundowning—when the elderly patients began to wonder aloud why they couldn't be home in their own beds. Those suffering from dementia or other cognitive decline got even more disoriented. It happened when the sky outside began to darken and the city's 7:00 p.m. applause wrapped up, with its attendant pot-banging and fire truck salute.

The patients couldn't understand why they were being asked to spend the night in this unfamiliar place. They complained of thirst or pain. They said they would never fall asleep with all the noise in the hallways and on the streets outside. They were lonely.

Mostly, they couldn't understand why their families weren't coming to visit.

Their cries and complaints echoed through the hospital halls. "Why didn't my husband come today?"

Gabriela reminded them each day that the hospital wasn't allowing visitors because of the pandemic, but they'd quickly forget. Every time she saw their bewildered expressions, Gabriela felt a fresh, prickly onset of pain, knowing how alone they must feel.

Even for someone new to elder care, the source of their confusion wasn't clinically hard to understand. The brain loses 2 percent of its weight for every decade a person ages past fifty. Its neurons, like the heart muscle, can't reproduce. The atrophy is most severe for the gyri, those ridges on the cerebral cortex responsible for so much of our thinking. Slowly the brain loses function, like a guitar shedding some strings and wearing down others.

"My daughter usually comes to see me every day. What's happened to her?"

"I know," Gabriela had to say each time. "I know how much they wanted to visit. But there's a very infectious virus right now, and we can't let them in the hospital. It's for your safety."

Then the patients stared back with accusatory eyes. Their requests for family visits would return the next evening, even more insistent, as though a son or daughter could be made to appear at the bedside by sheer force of will. Her patients who were hard of hearing and might have typically relied on lip reading strained to make out her words through the PPE.

The sundowning confusion was worse for those who didn't understand English and relied on phone interpreters, like Mr. Kaminski, one of Gabriela's patients who spoke mostly Polish. Each evening he demanded to know why he was still at the hospital. Mr. Kaminski, startlingly tall and equally thin, had broken his hip. This was one of the most common injuries Gabriela saw in her elderly patients—each year some three hundred thousand older Americans fracture their hips—but its familiarity made it no less excruciating. Mr. Kaminski was in searing pain. He filled the room with his intermittent cries. His groin and abdomen throbbed; his family was far away. He was on benzodiazepines, which just worsened his delirium.

His daughter worked in the medical field, so it was easier for Gabriela to explain his treatment when she called. But she could feel the daughter's fear from far away, how much she wished she could be with her dad. There was little worse than worrying for a family member you couldn't see. Especially someone who didn't speak English and couldn't communicate his own needs.

Gabriela was used to being the hospital worker that families loved, the one who spent ten extra minutes on the phone recommending the best times to visit. Now she could only parrot the rules, which were rigid: no visitors. An outsider was just one more possible vector.

"I'm keeping him company," she told Mr. Kaminski's daughter.

One afternoon Gabriela learned that Mr. Kaminski spoke at least one English word, when he grabbed her hand and cried out "Hi!" His lips turned up in a puckish smile. He still didn't understand this sterile white space, but at least he recognized Gabriela. "Hi!" he called out again whenever she came by. Gabriela liked to ask him questions about his hometown and his daughter, with an interpreter on the phone repeating her queries in Polish. It helped to distract him from his discomfort, though the conversations were clunky and filled with awkward pauses as they waited for the interpreter to make sense of their words.

His confusion lessened as they got his pain under control. "Soon you'll be ready for physical therapy," Gabriela told him. When the interpreter repeated this, his smile was evident even through his mask.

Another one of Gabriela's elderly patients had a variety of pains and problems, including an infection whose source wasn't clear. Her hemoglobin dropped one evening, and the nurse came to do a blood transfusion. She pressed her fingers along the woman's arm looking for a vein, then slipped in the needle. The patient winced. She grabbed Gabriela's hand and gripped it like a lifeline.

"I know it's hard, getting this thrown at you," Gabriela told her.

The woman nodded. "It hurts," she said. "I just wish . . ."

She trailed off. She looked over at the IV needle in her arm, with its attached plastic tubing and, above that, the plastic bag slowly pouring out blood.

"I wish I could have my family here," she continued, her voice so quiet that Gabriela had to strain to make it out.

"It's scary," Gabriela agreed. "It's hard doing this without your friends and family."

For a moment Gabriela felt like she was back on pediatrics. The woman sat there with her eyes closed, trying to steady her breath: inhale, exhale. Gabriela clutched her hand. The blood continued to flow. This was no different than Gabriela's youngest patients.

Alone at the hospital, without family, they were all like kids—bodies craving touch, craving certainty. Like they were learning to swim, letting go of the pool's side wall and kicking, kicking their legs as they looked for some familiar arm to hold. Finding no one, they gasped for breath.

Gabriela had plenty of experience caring for the elderly, if you included the hours she spent trading tense words on the phone with her family about Grammy's seizures and hospitalizations. But who could count that? The Covid wards were different. And it wasn't just the PPE and the worry of infection. It was the constant work of helping people prepare for the end of their lives.

In early May she transferred to an oncology team. One of her patients, Ms. Lau, had end-stage metastatic colon cancer. She was a quiet woman who almost never complained, though Gabriela knew she must have been feeling an ache in all her bones, the type of heaviness and discomfort that people feel with a bad case of the flu.

Ms. Lau was under five feet tall, and so slim that she looked as though she could be knocked over by even a forceful gust from an air conditioner. It was hard for the doctors to convince her to eat. The oncologist finally said there wasn't much more they could do. Typically, the doctors might have a goals-of-care talk with just one of her family members, but Ms. Lau's children all wanted to be involved. They scheduled a time to get the whole group on the phone: the patient, her children, the Chinese interpreter, the attending physician, and Gabriela.

"Do you know what it means when we say 'hospice care'?" the attending physician asked.

The phone interpreter then spoke, and Ms. Lau looked confused. She didn't seem to know what they were saying, though this wasn't a surprise. While Gabriela was preparing for the goals-of-care talk, someone had mentioned to her that the word *hospice* didn't have a direct translation into Cantonese; it could only be translated roughly to something like "deathbed care." Even in English its etymological evolution was more recent; the word used to refer to a rest house for travelers, before it took on its meaning for the elderly and terminally ill in the late nineteenth century.

The vocabulary of end-of-life care was especially difficult to translate because its implications were already confusing for native English speakers. What did it mean to be "comfortable," but dying? What did it mean to get care, but more for the purposes of pain management than recovery? These concepts were hard to untangle even if you knew their literal definitions.

"We're still going to be taking care of you, but we want to do what we can to minimize the pain you're feeling," the physician explained. "We want to make you comfortable."

She paused, as the phone interpreter translated this into Chinese. After several different interpreting attempts, the patient seemed to finally understand, and she told the attending that she was comfortable with the idea of end-of-life care. She wanted home hospice services, so she could die surrounded by her family.

But then her family members all began to speak at once, a heap of voices on the phone.

"Okay, thank you for this," one of Ms. Lau's children said. "We'll talk it over. We'll let you know what we decide."

Ms. Lau's children didn't seem to understand that it was their mother's decision to make, not theirs. Gabriela wondered if they hadn't fully heard or understood the conversation.

There was a cultural difference in play, too. In some Chinese families, medical care for elders was a family decision, one that could be determined without the patient weighing in. Sometimes a patient wasn't even told their diagnosis. But in American hospitals, if the elderly patient had capacity to make their own decisions, then the doctor had to honor those.

"It seems like your mother is comfortable with her decision," the attending said, waiting for the interpreter to pass the message along.

Gabriela looked to the patient, whose face was remarkably stoic. She thought about the strength it took to acknowledge that you only had a few months left with family. Fighting it took courage, but so did acceptance.

Gabriela was still exhausted the next day, which she had off from work. Her joints, eyes, bones—everything was sore. Jorge could tell. "Should we watch something?" he asked.

They scrolled through Amazon Prime, and Gabriela suggested the newish movie with the actress Awkwafina. "*The Farewell*?" Jorge said. "Let's do it."

They curled up on the couch, poured some wine, and started watching. It was the story of a Chinese-American family who learns their grandma has only months left to live. They didn't want to break the news to her, but they planned a family gathering to spend time together before she died.

Gabriela tried to keep her eyes fixed on the screen, but she was barely taking in the movie. Her mind glazed over with the image of Ms. Lau, and the sound of her children over the phone as their voices rose in protest. She thought of the woman's sweet, resigned face as they explained this foreign word *hospice*. Gabriela burst into tears.

"Are you okay?" Jorge asked, alarmed. They were only thirty minutes into the movie. "Should we turn it off?"

Gabriela shook her head. They stayed like that on the couch right through the movie's closing credits. Gabriela hiccuped and cried with a grief she hadn't felt so intensely since the last time she heard her own Grammy's voice from her hospital bed.

Fourteen

ELANA, May 2020

It was a Tuesday, around 8:30 a.m., and Elana was on the cardiac floor. The crackle of nerves that accompanied her earliest days had begun to dull. She felt more familiar with the protocols of Covid care—the temperature check downstairs, the sight of co-workers in their shields and gowns—and she didn't feel so jumpy when she introduced herself as the doctor. It was beginning to feel less like she was faking it.

"Morning, everyone." Elana's team gathered in a small work-room for table rounds. These were stressful moments that put a spotlight on residents who had to brief the attending on their patients' progress. The stream of questions that followed was always short, but sometimes grueling. Please let whatever's about to come out of my mouth be right, she thought each time.

Her phone kept buzzing, so at one point she looked down to check her lock screen. She had missed calls from her mom, sister, grandmother, and husband. She excused herself and stepped out into the hallway.

This time it was her sister, Daniela. "Dad's in the hospital," Daniela said. There was an uncharacteristic panic in her voice. "He went into cardiac arrest."

For the next hour, Elana felt as though she were watching her own body at a remove, seeing herself carry out all the necessary steps for crisis response without understanding how she was even moving her feet.

She told her team that there was a family emergency. Then she reached Akiva, who had been called when Elana wasn't answering her phone. He packed her a bag with clean clothes and a fresh mask while she drove home. She picked him up on their corner, drove to Long Island, and dropped him off at her family's home with her siblings. Then Elana went straight to the hospital where her dad had been admitted.

The hospital wasn't allowing any visitors, because of Covid, but they made an exception for Elana and her mother. Elana knew what this meant; her own hospital also made such exceptions. They thought her dad might be about to die.

She arrived at the hospital before her mom, and one of the nurses shepherded her to her dad's bed. He was motionless, eyes closed. Elana thought for a moment that he was already dead. Her heart began to hammer as she looked from the body to the nurse. Then she saw his chest rise and fall, ever so slightly. Elana learned that just before she arrived, he'd been resuscitated, and the doctors had started him on sedation, pain medication, blood pressure medication, and IV drips. They had also started him on cooling therapy, which helped to preserve the brain.

Elana's mom entered the room, a fresh wave of panic following after her, and ran to his bedside. "He's crying," her mother said. "Is he in pain?"

Elana looked over and saw her dad's cheeks wet with tears. "He probably can't feel anything right now," she said. "I think the tears are a reflex."

Elana's mother began to talk to her dad normally, as though they could have a conversation: "You shaved your beard this morning," she remarked. "You don't have a beard!"

Of course he couldn't respond. Elana forced herself to take a deep breath. If she broke down, she'd be of no use to her mother. Imagine he's a patient, Elana told herself. Not your father, but your patient.

Looking from her still, silent father to her mom, she considered what she might say if she was their physician. Patients like her dad who went into cardiac arrest at home had nearly a 90 percent chance of dying by the next morning. He was also at high risk for long-term anoxic injury, meaning he might not be able to think or behave like he had before the arrest. Elana wondered if he was already brain-dead. It was dizzying thinking about her father in clinical terms, though it helped that she'd gotten the call from family when she was rounding with her team, in doctor mode.

Six hours later the doctors ushered them out. There was a bed ready in the ICU, and Elana's dad would be moved over to critical care. The fellow on call gave Elana his cell phone number. "I'll call you if anything happens," he promised.

Elana got four calls that night. The first she slept through. The rest she picked up, as she lay in a hazy state of unease. Each call was a code. Her father went into cardiac arrest another four times. Every time Elana picked up the phone, she thought he might have died.

"Hello?" she said, dreading that voice on the other end.

But each time, he survived. By the time she got the 7:00 a.m. call that he had coded once more, she was ready to give in. Maybe three times he could survive, but a fourth? But the ICU fellow had a positive update: her father had made it through the night.

Theoretically, this was good news. But Elana was still convinced that he wasn't going to wake up. It felt like instead of marking her dad's death in that one moment, sitting by his hospital bed, she was mourning him hour by hour. A state of limitless grief. Maybe she would have to keep doing so for days, even weeks. It felt torturous.

Before the rest of the household woke up, Elana drove back to the hospital to get tested for Covid. The day before, she had gathered all her family members—mom, grandma, siblings—outside on the porch to discuss where she should stay to avoid exposing them to anything she could be carrying from the hospital. Elana suggested she could quarantine in the basement for fourteen days, but she could see from her siblings' stricken expressions that wasn't an option. Elana was the oldest and had always been like a deputy mom to the group. Her sister Daniela was visibly distressed, and her youngest brother, twelve years Elana's junior, seemed to alternate between shock and forced stoicism. Her brother Gidon, who was nineteen, was overseas and could only reach the family by FaceTime. Elana wouldn't be able to comfort everyone if she was isolated downstairs. She had called Montefiore to say that she would have to leave her own hospital team while she took care of family. She didn't feel conflicted at all about this decision, but she also wanted to make sure she'd actually be of use at home.

Elana called Sharon, her mentor, to ask what she should do so as not to put the rest of her family at risk. She couldn't expose her mom, or God forbid, her grandma to any stray virus she carried from the hospital, especially not with her dad already on a ventilator.

Sharon wouldn't sugarcoat; she never did. "Look," she told Elana. "You could always come and stay with me in my basement." This would mean Elana could see her family outdoors and at a distance but have somewhere isolated to sleep.

Elana felt a wave of gratitude for this maternal gesture. In the last day she'd quickly grown accustomed to being the source of parental care and authority.

"It's okay," Elana said. "I think I can figure this out."

She called the hospital and was told that as a health care worker she could set up an appointment to be tested for Covid right away. It took two days to get the results, during which Elana paced in the basement with Akiva, waiting for calls from the hospital and yelling words of comfort and encouragement toward her siblings up the stairs.

The morning that Elana's dad keeled over, he'd been fixing breakfast in the kitchen. Elana's grandmother, her mom's mom, heard a thump from the other room. She rushed in and found him on the floor. She was the only member of the household who knew CPR, but she wasn't strong enough to do compressions, so she screamed for someone to call 911. For a week after, they left his bowl of cereal untouched on the counter. Elana intermittently glanced over and contemplated cleaning up the mess, but it seemed too risky, like it could throw off some force in their world of lethally fragile balance.

From time to time, Elana wondered whether they really should have called for that ambulance. Maybe it wasn't right to have him resuscitated—four times, code after code. It seemed so unlikely he would make a full recovery. He would probably be dependent on a machine for oxygen. And it would be hard to decide when medical intervention had gone far enough; Jewish law was tricky when it came to cutting off life-sustaining treatments. A central tenet of Judaism was that almost any other religious law could be violated to save a person's life, as Elana's dad had reminded her when she was weighing working on the Sabbath: *pikuach nefesh*. This made it difficult to justify forgoing any life-extending medical measures. But what was done was done. They had called for the ambulance, they had given him CPR, and now his body was attached to a ventilator in the ICU.

Elana had never rotated through a critical care unit. She found that it helped her contain the tight coil of her emotions when she forced herself to continue imagining that her father was a patient she was studying, like she had in his hospital room. She took note of the lessons in care, treating each conversation like a study in the type of doctor that she one day could be. There was the attending physician who had trained at Montefiore and kept taking the time to ask about her career. There was the cardiac critical care fellow who insisted that she call him directly so she didn't have to maneuver through the hospital directory, negotiate with receptionists, and wait to be forwarded to the right extension. Each day he ended the call by telling her to "hang in there."

Elana used to worry, when calling patients' families, that she might be overstepping when she asked them how they were doing emotionally. If all she had to offer was pep-rallyisms—*You've got this!*—she thought maybe it was better to remain within the bounds of her own expertise, transmitting medical updates. Now she realized how soothing even a trite aphorism could be, a "Stay positive" from someone who knew her father's status and thought she should keep hoping regardless. In medical school, Elana was taught to foremost "consider the patient." But now that the patient was her father, she saw how much she needed the doctors to consider her, too. Every minute they spent on the phone with her was its own balm.

She took note of the clinical and biological lessons, too. Some ICU patients were on proton pump inhibitors, she learned the second day, which lessened the acid in their stomachs. She began reading compulsively about spontaneous cardiac arrest. *Roughly 90 percent of people who have out-of-hospital cardiac arrests don't survive. When looking at a cardiac patient's brain scan, the more definition you see, the better.* "This must be even harder for you, knowing all that you know," one of her dad's doctors told her. Elana thought this was probably true.

She wished that she could sit by his bed and talk to him. She had studied how family voices can help awaken the unconscious mind and speed recovery, by exercising the circuits in the brain responsible for long-term memory. But because of Covid, they couldn't be with him in person. Each day, she waited for one of the kindly nurses to FaceTime her, positioning the screen so she could see her dad.

Two weeks into his hospital stay, Elana's dad became medically stable. Then the question was whether he would wake up, and if he did, what neurological damage he might have. Would he ever be himself again?

Elana wasn't sure. Usually if she got this question from a patient's family member, she could sift through the pieces of clinical information and offer an answer with some degree of logic and clarity. But when she was talking about her own father, that logic was overshadowed by desire and a sense of hope untethered to reality. Did she really think her father would be himself again? Or did she just desperately want that, with every fiber of her being? When her mom asked her about his recovery, her mind turned more to a frenzied wish than to medical evidence. It was hard to give a straight answer.

Elana's father normally biked forty miles a day. He was an athlete. He ate nutritious foods, and he was strong and loved running around the yard with his kids. No one understood what had precipitated the cardiac arrest.

On the third week of his hospital stay, the cardiac fellow told Elana that if they didn't take out his breathing tube soon, they would have to give him a tracheostomy, which meant making an incision in the front of his neck and through to his windpipe. The doctors didn't say it explicitly, but Elana knew the downsides of this procedure: his recovery would be long and uncomfortable, he wouldn't be able to speak, and placing him in a rehab facility would be more challenging. Some rehab facilities didn't even take patients with trachs. The physician told Elana they were willing to extubate him experimentally, without a trach, but this bore a risk as well. He might not be able to cough or swallow effectively to clear out secretions, and his lungs could be vulnerable to infection. Elana's mother told her she should make the decision.

Going with the tracheostomy would be a reasonable choice, but Elana had the instinct that her dad didn't need it. She had spent hours that week on FaceTime with her dad, as an ICU nurse held the phone steady. Occasionally, she saw him try to swallow or cough through his tube. The doctors wouldn't have seen these small gestures because they moved in and out of his room too quickly. But to Elana, those motions indicated that her dad was trying to breathe on his own.

She spoke with the critical care fellow that afternoon, and told him that her dad seemed ready for extubation.

The doctor responded cautiously. "I want to make sure you fully understand the risks," he said.

Elana did know the risks. She knew that there was a chance that he could get pneumonia or be unable to keep his oxygen level stable. But the doctors were trained to make conservative estimations, and unlike Elana, they hadn't seen all of his body's subtle movements in the previous days. Also, their decisions were shaped by the clinical facts; Elana had to keep in mind the bigger picture of recovery and rehab.

"He can make it," Elana said.

Elana's stomach clenched as she hung up the phone. For the last four years, she had looked forward to graduating and bearing the full weight of medical responsibility over her patients. She'd been building toward the gravity of these intimate, high-stakes choices. Now she had that weight, but it was all coming too heavy and fast.

But the cardiac fellow called with good news the next day: "He made it."

"He's okay? He's safe?"

"He's off the breathing tube."

Elana thought, for a moment, about all the times that she'd been in the room while her medical team called someone to give life-altering news about their hospitalized family member, good or bad. The patient had survived a procedure. Something during surgery had gone awry. Elana had always tried to "treat every patient like a family member," but it was only now that she began to understand all that entailed. She was on the other end of that phone line, letting the news from a faceless doctor wash over her, watching the hours tick closer to tomorrow, because that would mean that he'd survived another day.

Fifteen

SAM, May 2020

As April turned to May, the city's high-flame panic tempered to a simmering dread. The subway began closing nightly for the first time in its history. Governor Cuomo announced that schools would not reopen before the end of the year. Fifteen children were hospitalized with a mysterious disease, similar to Kawasaki syndrome, thought to be linked to Covid-19. The city ran low on burial spots for bodies and started freezing them instead. Hospitalizations and new infections began to dip as the death toll rose, creeping above twenty thousand. At Bellevue, one of the unit nurses died of Covid. There was a memorial erected, a wall of cards and notes. Some staff members pinned ribbons to their ID badges.

Sam biked from home to Bellevue and back so many times that the turns began to feel reflexive. In his old life, pre-Covid times, the city used to pulse with commuting lifeblood. Now the streets were emptier, their flow quarantine-diluted. On First Avenue many of the people he passed were health care workers, headed to Bellevue or Tisch or the VA. It had morphed into some kind of trench, troops shuffling back and forth while the rest of the city slept.

One Monday morning, Sam took his usual route. The sky was cloudless, the sun bursting into a picturesque spring. Sam went south by Washington Square Park and up Lafayette. Then over to the bike lane on First Avenue and uptown to the hospital.

He locked his bike outside Bellevue, then went upstairs to his team. At 6:58, the night person got a page. A patient from Rikers Island needed a doctor to come see him right away so he could go to court, and he'd been assigned to Sam's care.

The relationship between Rikers and Bellevue was sometimes described as a revolving door. People who were jailed with serious illness were often sent to the hospital for treatment in its nineteenth-floor prison ward, though Rikers had its own small infirmary for less severe conditions. There was also a courtroom in Bellevue on the nineteenth floor, a wood-paneled room with stiff maroon chairs. But Sam didn't have much experience with it. He knew it mainly in the context of the forensic psych unit, which sometimes sent people to court if they refused their medications.

Sam went upstairs to meet the patient, who calmly explained his situation. His lawyer had told him that because of Covid, he was eligible for a possible release from jail. But to get the release he would have to leave the hospital, bus back to jail, and go to the court on Rikers Island.

In medicine they called this an AMA, "against medical advice." Typically, patients had to first prove that they had the mental capacity to leave AMA, which was determined using a four-part matrix called Appelbaum's criteria. This was taught with a mnemonic, CURA: the patient has to Communicate their choice, Understand the relevant information, display Reason in discussing the facts, and Appreciate the consequences. It was evident that Sam's patient checked all those boxes.

Sam wasn't sure how a Covid release from Rikers even worked. In the meantime, the patient wasn't anywhere near done with his medical care. He had an infection in his knee joint, which was actively destroying the joint. But Sam figured that being incarcerated would affect his health just as much as the infection, if not more. If this patient had the capacity to fix that, who were his doctors to stand in the way? Anyway, this man had just left one dysfunctional city institution, and Sam wasn't sure he needed someone from another institution, a moralizing "bad-cop doctor," telling him what to do.

The situation inside Rikers was, in the words of the jail's own head physician, a "public health disaster." By late April, 91 Rikers inmates out of every 1,000 had tested positive for coronavirus; for the rest of the city, the rate was 16 per 1,000. As pressure from the public mounted, Rikers was beginning to release those most at risk of contracting the virus—some 2,500 of its roughly 10,000 residents.

Sam called a pharmacy and instructed the patient to pick up antibiotics the second he got home, then come back to an emergency room, preferably Bellevue's. That wasn't easy for patients recently incarcerated, who got discharged back to the Bronx or Brooklyn and then had to make their way back to Manhattan, but it was helpful if they could come back to doctors who knew their history. With the paperwork and phone calls complete, the patient left to get the bus to Rikers.

That afternoon, Sam received another page from a nurse on his floor. The patient was back in the hospital's prison unit. He needed a transfer order to a civilian bed, since he was now free. It was unfortunate that he'd had to leave the hospital to get his release, but at least now he was back, and they could get him on antibiotics.

But once again, the patient explained that he had other plans. He wanted to go home, he told Sam. He planned, again, to leave AMA. This time, it was to see his kid and girlfriend. Then he needed to go to parole the next morning, and get his money back from Rikers, because his belongings had been seized when he was first arrested.

That Sam's patient prioritized his family members over his health after getting his jail release made sense—he likely hadn't been able to call or see them for weeks. Visiting someone you love at Rikers was hard enough in nonpandemic times; visitors often spent half a day just getting to the complex and through security, then had only an hour with their loved one in a loud, crowded room. But even these visits became impossible as the coronavirus began to spread. Jails locked down earlier than the rest of the city. Authorities made little effort to keep families informed. Partners and parents worried over their loved ones, with no capacity to contact them, and no understanding of what was being done to ensure their safety.

Who was Sam to tell this man he couldn't see his kid? Of course Sam also worried about whether in five years, or ten, this man would have healthy joints so he could walk and play with his son. This was the tug-of-war unique to somebody with one foot in the medical system and the other in the justice system. In the hospital, someone could monitor his leg's recovery and give him intravenous antibiotics. But coming from jail, he had larger concerns. Sam told him, candidly, that he couldn't relate to what was going on in the man's life. But of course he understood his priorities.

Bellevue's patients were a constant reminder of who really bore the pandemic's scars. While wealthy New Yorkers vacated the city or retreated indoors and bought up all the hand sanitizer, many of Sam's patients wrestled with more fundamental fears. The precariousness of their lives predated this global crisis. Their hardships were only amplified by these Covid weeks, as the question of whether they would be sickened by the virus became just another item on their list of uncertainties: Where to get food? Where to find shelter? It was hard to hoard toilet paper if they had difficulty affording it to begin with.

After he'd gotten the patient from Rikers another AMA, Sam's shift wrapped up for the day. He headed for the bike racks, ready to get home and have dinner. As he unlocked his bike, Sam looked up and saw a familiar face down the street. It was a family reunion, with someone on crutches: the patient from Rikers and his girlfriend.

The patient waved at Sam and introduced his girlfriend. "This is all her fault," he said. Sam wasn't sure how to interpret this, so he just smiled. How surreal, he thought, that less than twelve hours ago this man had been in jail. Now his girlfriend was picking him up to take him home. All in a single Bellevue day.

The same week, Sam had another patient sick with Covid-19 leave the hospital against medical advice. The reason: his landlord insisted that he had to pay rent. The patient had bacteria in his blood that could soon damage his heart valves, but his biggest worry was eviction. His health was already failing; he couldn't lose his home.

Sam had questions: *Can I write the landlord a letter? Can I fax him a letter? Can someone else pay your rent so you don't have to leave the hospital with bacteria in your blood? Do you have Venmo?* Sam knew there was an emergency city ordinance that offered renters protection in circumstances like these, but this patient was undocumented, and he certainly wasn't going to take his landlord to court. Adding to the complexity, the patient only spoke Spanish, and the landlord only spoke Chinese. The pandemic didn't overshadow the city's preexisting cruelties—it only put them in sharper relief.

Sam and his team often talked about Occam's razor, the principle that the simplest solution is often the right one. In the Covid era, this meant that everyone coming into Bellevue might best be viewed as a straightforward case of the virus. Shortness of breath? Coronavirus. Diarrhea? Coronavirus. Blood clot? Coronavirus.

Its medical counterpoint was Hickam's dictum, which argues that people can have many varying medical conditions, and identifying one doesn't rule out the possibility of others. It was coined by John Hickam, a Georgia physician famous for saying "Patients can have as many diseases as they damn well please." This was a somewhat less comforting notion; it implied that diagnosing a disease wasn't the end of the story, because every patient was likely a messy confluence of numerous problems.

At Bellevue—with its patients who were homeless, uninsured, and had underlying conditions like asthma—Hickam's words felt apt. Some patients admitted with Covid, like Sam's patient from Rikers, had so many other threats to their health and safety that the virus wasn't their primary concern. Many had ulcers or chronic obstructive pulmonary disorder. There were homeless patients who wanted to be discharged to sleep on the subway instead of in a shelter, because they knew they could isolate from the virus more easily on the A train than in a crowded city facility. There were homeless patients who preferred not to be discharged at all, because where would they sleep after leaving Bellevue? There were patients who couldn't get the organ transplants they needed because they were undocumented. These were the preexisting conditions of the American medical system: when the country's poorest turned up in emergency rooms, they were always already suffering from health problems that had nothing to do with the acute illness at hand, the effects of living without access to healthy food or clean water or consistent preventive care. Bellevue doctors called their main patient population the three "uns": undocumented, uninsured, and undomiciled (homeless).

In May, Sam admitted a medical mystery patient who brought all his ruminations about Occam's razor into sharper focus. The young man tested positive for Covid, but it wasn't clear whether his symptoms were caused by coronavirus or a different condition. He had diarrhea, and his kidneys weren't functioning properly. He didn't have breathing issues, but coronavirus was known to infect cells lining the large and small intestine, which could cause gastrointestinal symptoms. It could also cause acute kidney injury. He told the doctors that he had been feeling sick for two weeks.

"If your symptoms haven't changed," Sam asked, "what made you come into the hospital today?"

The patient explained thoughtfully that he'd waited fourteen days since the onset of his symptoms, because he had heard in the news that he was supposed to avoid the hospital if he could. Since he started at Bellevue, Sam had heard residents remarking that they weren't seeing as many bread-and-butter cases: GI bleeds, heart attacks, appendicitis. This didn't mean that New Yorkers were suddenly healthier. It just meant they were scared of the hospitals.

Sam was somewhat hoping that Occam's razor would prevail. The patient might simply have Covid, with an inflammation causing uncommon side effects. But it wasn't clear. With someone like this patient who came in reporting two weeks of symptoms, the temptation might be to simply say: Covid. But of course he was owed a better workup. There were all sorts of rare conditions that might explain his pain, which could coexist with the virus or result from it.

The young man's parents were scared for their son alone in the hospital. Sam called them with a Spanish interpreter on the line. It was difficult to explain all the questions that the doctors had about their son's case. Translating normal medical jargon to a family was hard enough; translating a doctor's list of uncertainties was even tougher. Sometimes Sam, with his limited Spanish, could pick up the interpreter's lapses in communication. "No no, we're not biopsying his lungs," Sam said, cutting in. "We're biopsying his kidneys."

The patient's mom was mostly worried about whether he had fresh T-shirts and a toothbrush. She couldn't come visit him, but the social worker arranged for her to drop a change of clothes in the lobby one evening. Meanwhile, her son didn't respond well to a blood transfusion. Immunosuppressive drugs might have helped him, but there was a concern that those could cause infection.

While his family waited for updates, this patient drew the attention of specialists across the hospital. He became a point of discussion at specialty rounds, the question still being: Were his symptoms Covid-related? Or were they signs of some other rare condition? He was mostly withdrawn, but his face took on a look that said: *Why is this bizarre thing happening to me?*

In medical school Sam had studied recency bias, the idea that recent experience with a certain diagnosis makes you more likely to make that diagnosis again in an unrelated patient. Sometimes his team talked about its variant, pandemic bias. Covid was everywhere: in the hospital emergency department, in headlines, in emails from family. It was hard not to let his mind land right away on that heuristic, even if the patients he saw might be coughing for some other reason.

Some of his classmates wondered aloud if being hazed on the front lines of coronavirus would warp their diagnostic vision, treating a stream of patients that were Covid, Covid, Covid, whereas no one person was just a single illness. Especially at Bellevue.

Sam was hot in all his layers of PPE. Beads of sweat pooled on his forehead and clouded his vision through the plastic face shield. Great, he thought. I'll just stand here looking like a goofy welder. Everything felt bizarre and harrowing all at once.

He was beginning to feel fluent in the protocols and vocabulary of Covid care. But he had just been assigned a patient who spoke only Cantonese. Sam's resident told him they should facilitate her goals-of-care conversation. She was terminal with cancer. Because of the restrictions on visitors during the pandemic, her husband would have to join remotely.

Sam filed into the woman's room on the allotted afternoon, trailing the palliative-care nurse. They dialed her husband and a Cantonese interpreter. As they were greeted by the voices on the phone, Sam wished that the patient's husband could be in the room for this weighty discussion, this confrontation with mortality. He wanted to see how the husband responded in real time. Normally the doctors mediating a goals-of-care conversation would reserve a room and gather with a patient's family members. They might leave at some point to give the family space to talk. Now the relatives were distant, disembodied.

Talking about goals of care in a foreign language brought another set of challenges, as Gabriela had seen. There was no type of conversation more important than one helping a dying patient determine how they wanted to die. Every word could impact the patient's emotional state and decision. Using an interpreter meant ceding some of that control. It also meant doubling or tripling the length of the conversation, as everything was repeated multiple times. (The worst was when the interpreters weren't good; Sam had seen patients slam down the interpreting phone and say "Try a different one," because too many words were getting lost in translation.)

But at least there was a loose script to follow. The health care providers began by asking patients to explain, in their own words, what they knew of their prospects for recovery. Often the patients vaguely understood already, but there was a power in articulation. It was the difference between hearing the string of clunky words that comprises a diagnosis and actually reckoning, in human-speak, with its implications. Guiding a patient through the maze of a terminal diagnosis wasn't easy, but it was well-trodden territory.

They started by asking the woman if she could tell them why she was in the hospital. Sam went out of his way to maintain eye contact with her. Using a translator phone made it easier to stare into the distance, because it could feel like you weren't talking directly to the patient. But this was too important for him to look away. He was still sweaty and uncomfortable, but he put his hands behind his back and willed himself to stand still without fidgeting.

Trying to remind her of the human behind the bulking layers, Sam reached out and took the woman's hand as they explained what home hospice care could look like for her. A hospice nurse would make regular visits and keep her comfortable. She would get to spend time with her husband.

"We're here to answer your questions," Sam said. Something he had learned to emphasize in palliative-care conversations was that the doctors weren't going to stop caring for you—the care would just look different. It wasn't about forgoing care, just reorienting the goals.

To Sam's surprise, the husband jumped in, asking if he could see his wife's medical records. Then he asked if there was someone else they recommend he talk to.

The palliative-care nurse practitioner, well versed in the choreography of goals of care, spoke next: "It sounds like you're asking for a second opinion."

This would be tricky in pandemic times. In a strained system like Bellevue's, it would be hard for the patient to get a new oncologist. The appointment would have to be by telehealth. Getting medical records in English from multiple institutions was already a nightmare—getting them in Chinese was even more complex. Anyway, Sam realized, her prognosis wasn't the view of just one doctor. If it was a fellow who had seen her originally, the fellow's supervisor had likely weighed in as well. Multiple eyes had looked at her case and come to the same conclusion: she was terminal.

But to his relief the patient cut in and told them she was ready to die at home. She didn't want to be nauseated and in pain as she fought through more invasive treatments. She wanted to spend time with her family.

Sam felt a sudden transference of gratitude to her. They were on the same page, sharing an understanding that her faraway husband couldn't. The oncologist had told them they had done everything they could for her, and anything else was not only exceedingly aggressive, there wasn't even consensus it was worth doing.

Sam set out helping the woman to fill in her hospice paperwork, what was called a MOLST (medical orders for life-sustaining treatment). There were stacks of papers with complex words and lots of places to sign, which she first needed to translate into Cantonese using an interpreter. Sam asked her if she knew how to read and write in Chinese, and she nodded, jotting down notes alongside each section.

Just before she left the hospital, she had to be tested for Covid. If she was positive, Sam had to coordinate her supply of PPE and warn the home health aide. Moving to home hospice was difficult, far more so with an infectious disease. How could she isolate from family in her last weeks spending time with them?

But one hour before her departure, Sam got the call: she was negative. He called the social worker, then the home hospice nurse. "She's negative," he said to each one, hearing their sighs of relief. It was almost like a round of telephone, points of information passed from one person to another, but at least for this moment it didn't feel so grim.

Sixteen

BEN, April 2020

Ben liked mixing cocktails. He could brew a homemade ginger beer with lemon peel, brown sugar, grated ginger, and carbonated water. He could make limoncello with just sugar, water, organic lemons, and Everclear. He typically tested out new recipes on his classmates, who knew he liked to entertain. During third year, on Christmas Day, Ben announced to his friends that he would be brining and roasting a holiday goose, which ended up being a nineteen-pound, $57 turkey, served with buckets of mulled wine.

Ben had practiced bartending one summer during college, after the trip to Sierra Leone he was supposed to go on got canceled because of the Ebola outbreak. He discovered that there was something about the repetitive nature of mixology that induced an immediate sense of calm. The measurements were easy and low-stakes, the outcomes predetermined.

Medicine he had always appreciated for the opposite reason—its alchemy of effort and the unexpected. His shifts in the emergency department during med school meant hours of unfamiliar faces and unpredictable events. In one night he might see strokes, chest pains, and gastrointestinal bleeding. He might see conditions he'd treated a dozen times before, and others that were entirely new. Ben enjoyed the fast pace, and the cycle of seeing a trauma, jumping to the requisite treatment, stabilizing the situation, and then moving on to the next patient. It was serendipity and trauma in uneven servings.

Right away, though, he sensed there wasn't anything that could have properly prepared him for the telemetry floor during the pandemic.

Ben's team in the telemetry unit, with its machines for monitoring heart rate and blood pressure, had taken in many of the most severe Covid patients.

Every morning their hospital was seeing five, six, seven codes. Before the pandemic, they would have seen five or so codes a week; at the Covid peak they hit thirty per day. Some of the patients survived and were transferred straight to critical care. Others died on the spot. Some of them were names and faces Ben knew, having already taken their medical histories, but others he met for the first time as his team attempted resuscitation.

One of Ben's co-interns had a patient, someone he discussed during rounds, who had renal failure and end-stage heart failure. Ms. Henry was a tiny woman whose chances of making it out of the hospital seemed slim. But her family wanted her to be full code—intubated and resuscitated in case her heart or lungs failed.

"Do everything," they told the team. "We just want her to live."

One morning Ms. Henry had to go downstairs for dialysis. Ben's co-intern called the transport team, which was nerve-racking. When a patient left their room for treatment on a different floor, their care was (however briefly) out of the team's hands.

A few minutes later, Ben heard a voice over the loudspeaker calling a code. He was so used to reacting right away when he heard that voice that he jumped up and started moving before his brain processed that the location was in dialysis. Someone on Ben's team called downstairs to ask who the code was for. It was Ms. Henry.

"Where's dialysis?" one of his team members asked aloud, and they scrambled to look up the floor while sprinting toward the stairwell, extra PPE in hand in case there wasn't enough on the lower floor. Masks, gowns, and gloves were vital for chest compressions, when the doctors had to lean close to a patient's face, and especially for intubations.

When they arrived, Ms. Henry was pulseless. The code and support teams clustered around the gurney in the dialysis hallway. Someone looked up her chart as Ben and another resident started compressions. They rotated every two minutes, with a pause to check if Ms. Henry's pulse had returned.

Ben's heart pounded wildly as he pressed his hands into her chest. There were sounds all around him: monitors beeping, team members shouting "Check her pupils!" It was typically the third-year residents who led the codes, with an attending physician supervising nearby. The fellow who was leading the code took on a tone of unwavering authority, a just-trust-me voice that counteracted the unease bred by the surrounding commotion.

As Ben pressed downward, he heard again the Hippocratic oath echo. *Do no harm.* It was a promise that could feel strangely incongruent with the aggressive, sometimes futile process of resuscitation. He thought about her family's instructions, too: do everything.

After thirty minutes, the team called the code. "Time of death?" someone asked.

Ben felt his own chest squeeze tight. He didn't know much about Ms. Henry and her personal life, because his co-intern had been responsible for her care. But the sense of loss was still all-encompassing: this patient, whose children had so recently prayed for her and made hopeful plans for her release, was now gone. Soon her family would receive the news in a phone call from his co-intern. And Ben's role would be to shelve his feelings away for now, keeping his eyes on a to-do list bristling with the needs of his other patients.

Ben turned to leave the scene. As he looked down, he saw a streak of red—the dead woman's blood on his scrubs.

Along with the Coalition Forces of early graduates came the Allied Forces. These were doctors deployed from nonemergency and internal medicine units across the Montefiore system to help out as the crisis mounted. There were orthopedic residents, gynecologists, dermatologists, pediatricians. But they were all treating Covid patients now.

One of Ben's teammates was a psychiatry resident who dutifully transferred to the cardiac telemetry unit when the pandemic struck. Sometimes she looked to Ben for help on the more clinical tasks—drawing blood, getting patients on IV drips.

He also stepped in for arterial blood draws whenever it was tough to hit a vein. The process was straightforward, but it took the sort of recent practice someone in psychiatry was less likely to have: grab the radial artery, feel for the pulse, insert a special needle that's held like a dart. They were also more painful than typical blood draws.

It felt odd to Ben, stepping in where the second-year resident couldn't. The medical trail that he had learned to navigate placed value on seniority. Every time you edged up the ranks, you got someone just below you watching your lead. Here was someone with one more year of training than Ben—yet she was reliant on his steady hands.

But Ben liked to watch the Allied Forces at work. They didn't always have the same clinical fluencies he had, but they had a different orientation to their work. The psych resident had a way of speaking to families on the phone. She seemed practiced in the vocal cues that eased the friction in these conversations.

Another one of Ben's new colleagues had been redeployed from orthopedics. She was on a different floor, but he saw her sometimes around the building during a lunch break. He had first met her years ago, during a late-night shift in the emergency department when he was a medical student. They'd been treating a young boy who fell off a skateboard and broke his wrist. Ben helped the resident do a reduction to set the bone and put the grateful patient in a cast.

The orthopedics resident told Ben that she sometimes looked longingly over the console to see what fractures were coming into the ER. All she wanted was another broken leg. Ben, too, found he missed the ease of broken bones. They were spending their days running codes, attending to conditions that couldn't be fixed by anything as straightforward as a cast.

Ben felt a particular affinity for his senior resident, with whom he worked closely in determining plans for each patient's care. The resident was headed for a critical care fellowship at Montefiore. She was obviously drawn to the teaching part of medicine. She grew animated when interns asked her about the basics: What do you do for someone with chest pain? How do you decide when to give someone a catheter?

Watching her, Ben thought about the type of medical teaching he might like to do someday. His ideal would be working with the third- and fourth-years, who were beginning to settle on their chosen fields but were still wide-eyed about other specialties too. He'd have to complete a fellowship specifically focused on teaching, and then he'd get to train medical students during their clerkships and help familiarize them with the essentials of emergency medicine.

One morning, not long after he'd started work, Ben heard from a coworker that a Montefiore resident was sick in the ICU with Covid. The rumors trickled through the hospital floors. He was a third-year in the emergency medicine residency. Ben recognized the name but hadn't met this young doctor. Slight with thick black glasses and a shock of dark hair, he called himself a "badass emergency doctor" and typically bounded into the emergency room in a burst of energy. He was also known for soldiering through late-night shifts with a fortitude that didn't seem to fray. But Covid had brought him to his knees. When he came to the ER as a patient, he was mortified by the idea of relying on a bedpan and watching a nurse wipe his bottom. The nurses he usually worked with asked him what he was doing there, and he told them that he'd tried to stay home. He hadn't wanted to add to their patient load.

Soon afterward, the doctor's fight for his life was written up in the *New York Times*. He was twenty-seven. Exactly Ben's age. He was put on a high-flow nasal cannula. Supposedly one of the doctors who saw him admitted, and also ran the residency program, later went into his office, put his head down on the desk, and cried.

When Ben shared stories from work with his roommate Sean, the conversations could feel surreal. Sean, who would soon go to Texas for residency, had been put on a Covid team at Montefiore Moses that so quickly ran out of patients it was converted to normal medicine. It could feel like he and Ben had been sent to work in two different cities. But it was just luck. No two teams were alike. For many, the number of Covid cases had quickly plummeted, and they were easing back to the norm of heart attacks and appendicitis. For Ben, on telemetry, the stream of frightening virus cases didn't stop.

The brightest moments in Ben's week came when the hospital halls rang with Jay-Z's "Empire State of Mind," which meant a Covid patient was being extubated or discharged. Each hospital had its own soundtrack. Lenox Hill played the Beatles' "Here Comes the Sun" (with nurses calling "Code sun! Code sun!"). Hackensack University Medical Center played Bill Withers's

"Lean on Me," Maimonides in Brooklyn rotated hits like U2's "Beautiful Day" and the "Rocky" theme, and NYC Health + Hospitals Metropolitan in East Harlem chose the Journey classic "Don't Stop Believin'."

But these celebrations were fleeting. The death toll on Ben's floor kept rising.

In early May, Ben had a patient admitted with Covid who seemed to get sicker by the day. Mr. Wilson was in his eighties. His oxygen saturation kept declining, meaning his needs would outstrip the ten liters of oxygen that his nasal cannula could provide. His creatinine levels were also high, which signaled that his kidneys were failing. Ben had to call Mr. Wilson's kids to explain that their dad had end-stage kidney failure.

"He's going to need urgent dialysis," Ben told them.

It wasn't until Ben began recommending dialysis that he fully understood how it altered the course of a patient's life. The easy view of the treatment was as some miraculous intervention—which it was. It could keep the body in balance by removing waste, salt, and extra water, and maintain safe levels of chemicals like potassium, sodium, and bicarbonate in the blood.

But it often meant patients had to restructure their lives around their disease. Those with chronic kidney disease had to visit a dialysis center three times weekly for four hours, or spend hours each night with a massive machine hooked up to their abdomen pumping fluid. It meant completely rewiring their relationship with food. The dialysis diet entailed strictly limiting intake of potassium and phosphorus, since the kidneys couldn't manage electrolytes. Some of Ben's dialysis patients told him all they wanted was a banana. In school, during a unit on the kidneys, his professor said one of her dialysis patients once desperately wanted to eat pasta with red sauce on her birthday. She went into cardiac arrest right there in the restaurant; the potassium load from the tomatoes had stopped her heart.

Ben had to explain all this to the anxious family members who just wanted reassurance that their loved one would be safe. On the phone with Mr. Wilson's children, he groped for the right words. Their dad's odds of surviving his hospital stay were slim. Ben wanted to know if the kids would be open to comfort measures so he could focus on minimizing Mr. Wilson's pain levels.

But Mr. Wilson's son walked a hard line. "We want you to do everything you possibly can," he told Ben.

After a pause, he added: "Our dad is a fighter."

This was one of those goals of care terms that made Ben uneasy. *He's a fighter.* But does he know what he's fighting? Ben wondered. Their dad's mental state had begun to deteriorate—because his kidneys weren't working, toxins were accumulating in his brain that left him confused and disoriented. He was in hypoxic respiratory failure, meaning there wasn't enough oxygen in his blood. He was too sick to go down to the dialysis unit, so he would have to get dialysis at his bedside.

But Mr. Wilson's son seemed to have made up his mind even before the call. "He's been through a lot. I know he'll pull through this."

For Montefiore's sickest patients, and those with especially prolonged hospitalizations, the hospital had made arrangements so family members could come visit, though they had to have their temperatures screened, wear masks, and limit their in-person time with the patient. A few days later, Ben ran into Mr. Wilson's son, exiting the elevator onto the eighth floor.

"I want you to know," Ben said cautiously, "your dad is a little bit out of it right now. He'll probably seem confused to you, like he doesn't know where he is."

The son's face was drawn in worry. "I got it," he said. "I understand." His confidence from their phone conversation seemed to be dissipating. Like the new doctors, families were processing how totalizing the virus could be in its assault. They were all confronting its cruelties in real time.

Mr. Wilson died not long after, during a code.

Do everything you can. That was the soundtrack to Ben's days: phone calls to families who insisted their loved ones could somehow survive. They wanted their mom or sister or partner to be full code. Ben kept stumbling over the phrases people used when they talked about these preferences. It was the same few words: "Do everything." "We want her to live." It seemed to brush aside some of the fibers of complexity that texture a medical dilemma, namely: What kind of life do you want her to live? Because even if she does live, her life won't look anything like the one she led before.

These voices seemed to think about medicine as formulated around the doctor-as-hero, the white-coat savior who can fix a kidney as easily as a shattered wrist. Ben was more interested in the doctor-as-mediator, helping people determine what parts of their lives they actually wanted to save. This meant conversations were more muddled—an effort to distinguish between "life" and "a good life," and uncomfortable decisions when the inability to have the latter meant the former was no longer worth pursuing. But it wasn't Ben's job to persuade. He was just a doctor, an intern, and not a philosopher king.

There was a flow of patients entering the hospitals whose cases were progressing so rapidly that there wasn't time to talk with them about their end-of-life wishes. Patients were showing up in the emergency room with hypoxic respiratory failure who were intubated within minutes. There were patients who, even after intubation, required invasive procedures in order to live. They would need to spend weeks on ventilators. Some would spend a full month in the ICU. Even after progressing to a step-down unit, they might suddenly code.

In the ideal world, Ben thought, these patients would have had their goals-of-care conversations with family members long before they turned up in a pandemic ICU. But kids were understandably reticent to ask their aging parents how they wanted to die. All the research that Ben read showed the lengths people would go to in order to avoid discussing death. Ben recalled the moment of silence on the phone when he asked his own sister if she could be his health care proxy. No one wanted to be reminded of the mortality of someone they love.

By the start of Ben's fourth week, the conversations still weren't easy, but they began, at least, to have a frictionless familiarity. Like drawing blood or suturing a wound, their facilitation was a skill that was fortified with repetition. Ben had learned to explain to anxious families how the virus tore through the body, and the damage it left in its wake.

Take hypoxic respiratory failure: "The virus is attacking her lungs, so her lungs can't get oxygen to the blood," Ben would explain. "So what we're doing is we're giving her lots of oxygen to try to maintain the levels of oxygen in the blood."

And if the patient worsened: "The oxygen levels we need to give her keep increasing. Soon we'll max out on what we can give her through noninvasive measures like a mask. That's when we'll have to decide if we want to intubate, or put a tube down her throat."

Why wouldn't we, if that could possibly save her? "Intubating can do serious damage to the vocal cords long-term. It can cause its own infections, or even puncture the tissue in her chest cavity, which could cause her lungs to collapse." The weaker or older the patient, the more severe the risks. And once someone was put on a ventilator, it was immensely challenging to wean them off. Because the ventilator would be breathing for them, their diaphragm could quickly lose its capacity to function properly. One doctor metaphorically described this as doing pull-ups for weeks with full body weight assistance, then suddenly taking away all that support—the person would just hang there.

Ben could share numbers too, about the chances of survival if someone was intubated, but often those statistics crumpled at the feet of a family member's optimism. Families clung to rosy images of their loved ones, the reality of failing organs erased by a stubborn nostalgia. Ben remembered once speaking to a patient's wife, who swore that before her husband was admitted to the hospital he had been perfectly healthy, capable of making eye contact and speaking lucidly. But when Ben called the nursing home, the staff said the man usually stared off into the distance and didn't even respond to his own name.

Sometimes Ben had to be blunt: "I don't think compressions will make a difference. I don't think shocking will make a difference either. I think that'll just prolong the dying process."

The voices Ben heard in response differed with every call, but he learned the types of responses that helped. Sometimes he let the family members sit in silence, so they had time to grieve. This wasn't the grief that accompanies death; it came from the realization that even survival would entail life-altering consequences. Ben learned to pose certain questions: "What would you like from me?" "How can I help right now?" "What other information can I provide?"

He heard other cries that bordered on denial, from adult children who asked him: "Can you make sure you're doing everything possible to save her life?" Of course they wanted intubation, of course they wanted resuscitation. And yes, they understood all the risks.

The calls with families felt like an injection of acute emotion. Ben could spend hours tending to someone barely conscious, only coming by their bed for minutes to check on their vital signs. Then on the phone with family, he learned all the details that colored their lives outside the hospital. Some were teachers or firefighters, or had bartended at their neighborhood joint for decades.

One of Ben's patients was exposed to coronavirus at a nearby nursing home. She came to the hospital with Covid and signs of kidney failure.

Ben called up her children and explained that there was little his team could do. "Right now she's delirious," he explained.

When the kidneys failed, the brain sometimes went with them because the electrolytes in the body were off balance. Toxins accumulated, causing lethargy, confusion, and even seizures.

Ben worried that this woman's children would insist she keep fighting. At this point, he thought her best option was to minimize suffering. Intubation didn't seem likely to save her. But the kids were understanding. "We want you to do whatever you can to keep her comfortable," they said. He was relieved when they told him to mark her as DNR and DNI. In her stage—elderly, already in failing health—there was little that could be done in these final days but to numb the pain.

Once in a while, Ben considered how he might respond if he was in a situation of life-threatening illness. He liked to think that if he was diagnosed with some incurable disease, his response wouldn't be of the "Oh my God, do everything" variety. But it was hard to say how he'd really feel. Following medical logic and accepting foregone conclusions seemed doable, until it was you and your family on the tough end of the prognosis. Then it all began to unravel.

The doctor's tense relationship with death—doing everything to ward it off, but sometimes inadvertently speeding it along—is as old as the profession itself.

For three thousand years bloodletting was thought a powerful answer to most ailments—smallpox, epilepsy, gout, pneumonia, cancer, even a difficult delivery. The standard response was to nick someone's vein and let the blood flow to rebalance what Hippocrates called the four humors: blood, phlegm, black bile, yellow bile. It's now understood that in some cases this likely accelerated death, leaving the body in low blood pressure and shock.

Bloodletting was an early sign of the field's core tenet: mortality can always be challenged. As a result, doctors have long tended toward aggressive intervention. Only in recent decades have they begun to recognize when these invasive measures hurt more than they help.

When the AIDS crisis reached New York, these conversations were not standard practice. Some patients, who fell sick after already having seen their friends tormented by the disease, had no interest in life-sustaining measures that would prolong their lives only by a paltry, painful few months. But hospitals didn't have clear procedures for communicating these wishes upfront.

The doctors were also aware that many of their patients were staving off the inevitable. Sometimes they were said to do a "slow code" for an AIDS patient. That meant responding to a heart or lung failure without the speed and urgency needed to actually save the patient. "This wasn't cowboy medicine," one Bellevue doctor said. "Nobody tried to play God. These people were unbelievably sick, they had no chance of getting better, and they were going to keep suffering until the minute they died."

Out of this period came the impetus for more meaningful conversations on preparing for death. Thomas Wirth, a gay artist from Greenwich Village, came to Bellevue in the 1980s with HIV and toxoplasmosis (a brain infection). He had seen loved ones grow emaciated and agonized through their final months fighting AIDS. He left a note for a friend, John Evans, with written instructions to withhold "life-sustaining procedures" when it looked like he no longer had a chance at meaningful recovery. And he gave Evans the power of attorney.

But when Wirth fell into a coma, the hospital sidestepped the AIDS diagnosis and determined that his brain infection might not be lethal, so they could not withhold treatment. This led to an influential lawsuit on end of life care, *Evans v. Bellevue*. (Bellevue won the case, but the patient died the next month.) In 1985 New York governor Mario Cuomo formed the New York State Task Force on Life and the Law, which endorsed the ideas of a living will and designated proxy so patients could record their wishes before they got too sick.

Through patients like Wirth, the AIDS crisis began to reshape the idea of palliative care, and with it the medical field as a whole. It forced a reckoning with the purpose of the profession: Do doctors have an obligation to keep fighting a disease just because they can? Who has the ultimate say over life and death—the patient or the provider? Whose hands are controlling the clock?

Nearly four decades later, the Covid-19 pandemic has brought with it hundreds of thousands of these difficult discussions. Patients who hadn't given their end-of-life wishes much thought were suddenly prompted to articulate their desires. Families unfamiliar with these concepts were forced to learn the language.

As the crisis mounted, some doctors observed with wonder that patients and families were proactively *requesting* goals-of-care conversations, even when their medical conditions weren't life-threatening. Physicians are accustomed to delicately raising the prospect, knowing the worry they'll spark in doing so. But Covid reshaped the context. Families had grown aware that a case of Covid could turn fatal without much warning. They didn't want to risk having someone fall critically ill or unconscious without first stating their medical preferences. As Ben saw firsthand, the questions were too important and complex to wade into at the moment when someone's lungs were failing.

The pandemic also brought new wrinkles to these conversations. Families could not be in the room when patients were discussing goals of care. Physicians could not use physical touch to comfort their patients. They couldn't even use their facial expressions to communicate, given their PPE. They had to unlearn all the things they'd been told: *Never rush the conversation. Sit down by the bed instead of towering over the patient.* That unlearning could be especially hard for the interns, like Ben, for whom the lessons themselves were still fresh and raw.

Some of these challenges weren't new to the pandemic. Working-class patients whose family members cannot afford to take off a midday shift have often had to articulate their goals of care with their loved ones on the phone instead of seated side by side. Patients who do not speak English have always found end-of-life conversations more taxing, because the subject is not only emotional but also full of unfamiliar vocabulary. In making these difficulties pervasive, the pandemic exposed how much harder these conversations had been for the marginalized all along.

Dr. Gawande put it this way in an interview: "The pandemic has made people across the age spectrum realize that they are vulnerable, that people are frail, that you are a mortal being. We are asking ourselves, what's the quality of life I'm willing to consider acceptable, and not acceptable, in order to preserve my survival?"

For much of the last century, doctors seemed to define the success metric of their work as survival, not comfort. Their job was to outsmart the natural forces of aging and disease, and they often did it well. With every new scientific advancement, that power expanded. Conditions that were once death sentences became just fancier puzzles. And this made the divide between the average doctor and the average person seem profound. The role of medicine was to do what the rest of us couldn't: dodge death. But in crises like Covid, the divide comes undone. Patients spiral quickly, and doctors know they aren't always in control. Sometimes the goal can't be to avoid or fight off terminal illness, but just to ease the confrontation.

Perhaps it helps that, since March, people have become more accustomed to thinking in trade-offs for all decisions, not just high-stakes medical ones. Going to the grocery store means accepting some level of danger. So does dining indoors, or attending a wedding. The architecture of routine is suddenly made of risk and choice, weighing quality of life against certainty of survival. Which means that doctors and terminal patients are less alone in their tangling with mortality.

Seventeen

JAY, May 2020

Jay felt a gnawing sense of distress when it came time for her phone calls to a young man whose parents were both in the hospital with Covid. The father, Mr. Wang, was in the ICU with the disease. The mother, Mrs. Wang, was Jay's patient in the step-down unit. The son's voice on the phone was cracked and worn whenever he and Jay checked in.

Mrs. Wang, a dark-eyed woman with long gray hair parted on the side, was faring better than her spouse. She was on ten liters of oxygen from a nonrebreather mask. Her husband, meanwhile, was intubated and unresponsive. Mr. Wang had suffered a middle cerebral artery stroke while in the ICU and been put on medication that caused temporary paralysis, which helped him to tolerate intubation. His health care proxy was his wife, but she was too medically unstable to consult for the time being.

"Can I ask you a favor?" their son asked one afternoon.

"Of course," Jay said. "How can I help?"

"I know this might not be an ideal time for this, but my mom keeps asking if she can Zoom with my dad," the son said. "She's worried about him, and I think it would help if she could see him on video."

Jay swallowed. The son hadn't seen his dad in over a week, so maybe he didn't fully comprehend: his dad couldn't breathe on his own, let alone open his eyes or formulate words for a video conference screen.

"Your dad isn't in a place to talk or respond," Jay said. "So I'm not sure if a Zoom would be helpful to your mom. I think it might upset her."

The son had been told that his dad was unresponsive, but there was a difference in hearing an ICU report and accepting it. "Oh, right," he said. "I knew that. I guess I just had to ask for my mom, you know?"

The next day, Mrs. Wang transitioned from ten liters of oxygen to five. It was good news that also suggested she was moving toward a place where she could advise on her husband's care. Jay visited Mr. Wang in the ICU. He was still nonresponsive, though they weren't sure whether this was due to his stroke alone or to Covid encephalopathy in addition to the stroke. It was time to discuss his priorities and goals of care, and the palliative-care team asked Jay to convene a meeting for Mrs. Wang, her son, her husband's ICU doctors, and a Mandarin-speaking attending physician.

As Jay was about to enter Mrs. Wang's room, the attending shot over a stern look. "You're not going in there," she said.

"What? Why?" Jay asked.

"She's still Covid-positive," the attending physician said. "I'm interpreting, so I have to be there in person." Many patients did rely on interpreters joining by video, but it was much easier to hear what they were saying in person, especially when they were on supplementary oxygen. "You can join by Zoom."

Jay opened her mouth, getting ready to formulate some protest, but the attending shook her head.

"I'm not going to let you in the room," she said, in a tone as welcoming of dissent as an autocratic regime.

Watching her attending disappear into the room, Jay made a mental note: This was the kind of mentor she wanted to be to a resident someday, protective in a nonpatronizing way. Jay felt a pinch of guilt, but an even more potent sense of relief. Mrs. Wang was still contagious. Every moment at her bedside meant more risk of exposure. Jay's fear of infection from the earliest days on the wards had softened, but it hadn't vanished, especially because she was at heightened risk with her asthma.

Jay stood right outside the door to Mrs. Wang's room, peeking through the window, then pulled up a computer and clicked into the video meeting. She saw the team clustered by Mrs. Wang, who was fighting to be heard, though each time her voice rose she broke into a coughing fit. Someone suggested she try writing instead of speaking, which proved slightly more effective, so they began passing notes back and forth. The ICU doctor wanted to discuss Mr. Wang's care moving forward, and especially whether Mrs. Wang and her son supported giving Mr. Wang a tracheostomy.

"I've actually talked about this with my dad before," the son said into the Zoom camera. His voice was wobbling. "We've discussed it many times."

The palliative-care doctor looked over to Mrs. Wang for her guidance. As the health care proxy it was ultimately her decision, but clearly the son, her eldest, had given it significant thought, and Mrs. Wang seemed to defer to him.

"I'm worried about his quality of life with a trach," the son said. "Personally, I know I wouldn't want to be dependent on a machine to live."

"That's understandable."

"But my dad said something different."

"Yeah?"

The son explained that his dad wanted anything done to ensure his survival. If that meant a tracheostomy, then he would want a tracheostomy. Then he broke down crying.

These conversations were impossible from the outset. No kid liked acknowledging a parent's mortality. But there was an extra layer of difficulty when one family member disagreed with the other. Some people just wanted the bare minimum, survival. But others wanted more—to dress themselves, feed themselves, walk to the corner grocery store, hug their grandchildren. Jay could imagine one thousand definitions of a fulfilling life, each one made to seem less realistic by her team's diagnosis.

Jay listened as the son explained that he and his dad had argued repeatedly about end-of-life care. His dad had maintained that he wanted any intervention, regardless of slim prospects and cumbersome rehab. It was gut-wrenching hearing him make the case for his dad's medical wishes while fully understanding how painful those would be to enact.

Jay ached for the son, watching his parents fight the same disease. Rationally speaking, it made sense that the virus tended to sicken whole families. Of course its particles could easily circulate through a household. But it also made for particularly agonizing patient clusters, a knot of ailing parents and worried children.

Back at home, Jay missed her own family. As relationships go, theirs could be thorny. They fed her insecurities; their worries became hers. Sometimes her head rang with her dad's warning all those years ago, telling her she would quit medicine ("That's not women's work"). But then she got her mom and dad's phone calls and heard their voices textured with concern. Like Mr. and Mrs. Wang's son, all she wanted was to have her parents back, physically close. She was watching a son fight for his father's life. It was hard to feel like anything else mattered.

Mrs. Wang was discharged to a rehab facility shortly after that conversation, but her husband remained ventilated in the ICU. Meanwhile, Jay was spending her evenings at home worrying about her own mother back in Westchester. She could imagine her mom's loneliness, with her kids either on the other coast or close by but unable to visit.

"Can't I just come for a short visit?" her mom asked one evening on the phone. "I can drive into the city. We'll keep six feet apart."

"We can't, Mom," Jay said. "It's too dangerous. I've been exposed in the hospital every day." This was the problem with being the only doctor in her family. Jay was not only most at risk, but also most aware of the risk she could pose. She was both carrier and enforcer.

"You know, I am the one who birthed you," her mom said, only half joking. "I should have some say in all of this, don't you think?"

Kate seemed to understand the logic of staying away from her daughter, on good days. But sometimes her maternal reflexes overwhelmed her pandemic reasoning. She couldn't adjust to being less than an hour from her daughter by car but unable to see her. Especially when the world seemed to be falling apart. When everything comes undone, hold your child; that was her instinct, programmed so deeply that it couldn't be shut off by rational argument.

Jay called the day before Mother's Day. "I have a present for you," she said. "But I don't know what to do with it. I want to make sure to keep you safe."

"All I want for Mother's Day is to see you," Kate replied.

"You know I can't see you in person."

"You know what?" her mom said. "I gave birth to you. I love you. It's Mother's Day, and if I want to see you, I'm going to see you."

Jay paused. "Would you feel okay with talking through a car window?"

Kate drove into Manhattan the next day and pulled up outside Jay's dormitory building. Jay stood on the curb, waving as her mom parallel-parked. Then Kate called Jay on the phone and they sat there giggling for two hours with the car windows rolled up. Jay tried not to talk about her work at the hospital too much.

This exchange, eyeing each other through the glass, seemed to float outside their complicated moment.

"This was all I wanted," Kate said. "This is perfect."

As her mom pulled away to return home, Jay realized that she hadn't been hugged by her mom since early March. She could feel the distance between them like an overwhelming force. She had clutched Manny's hand, and felt his cheek resting on her arm as he napped. But she missed unconditional, unworried touch. Still, it satisfied some itch she hadn't even realized she'd been feeling just to see her mom in person, instead of on a FaceTime screen.

Jay's brothers, Christian and Jackson, were living in California. The boys were characteristically cavalier as Santa Clara shuttered around them. Jackson was excited that his college classes were now online, because he could get through the recorded lectures at twice the speed. Christian had been hired as a physical therapist at a clinic, which announced it was shutting down the day before he was meant to start. Now he was getting an online certificate in sports medicine instead. The ease of their quarantine lives—sleeping in, longboarding to the grocery store—was a stark contrast to Jay's hospital hours. The three of them were close, having been raised without cable television and trained to rely on one another for entertainment. But now their conversations were strained by the distance and the difference in their day-to-day obligations.

Still Jay's overprotective sister instincts kicked in over the phone. "Are you guys wearing masks?" she asked them when they FaceTimed. "Are you being careful?"

Jackson called one afternoon, and the minute Jay picked up and heard his voice, she could tell something was wrong. Jackson had been calm throughout the last weeks, directing most of his stress toward his online classes. Now his voice was quivering. His roommate had tested positive. "What if I have it?" he asked Jay. "What should I do?"

The return to their typical roles was comforting. Hearing him pace the house and try to lower his pitch brought back flashes of her little brother tumbling into the snow or falling into the pool. Jay would always be his de facto doctor, even from the other side of the country.

"Plenty of people your age are asymptomatic," she told him. "If you're not feeling anything now, the most important thing you can do is stay home."

Jay wasn't sure it made sense for him to longboard over to a testing site, where he would be surrounded by coughing people who likely had the virus. She wasn't surprised, though, when he texted her two hours later to say he was at a clinic getting swabbed, just to be safe. She could only do so much in her long-distance maternal role.

Jay darted down the hallway with a damp blue sponge in her hand. She had noticed that morning that one of her patients, Mr. Wu, kept licking his lips for moisture. He wasn't the type to complain to the nurses about the dehydration, nor could he speak much English. But when Jay appeared by his bed with the sponge in hand, he grabbed it delightedly and pressed it onto his chapped lips. His water intake was restricted because he had a heart failure exacerbation, and the water would've backed up his lungs, making it harder to breathe. The moisture on his mouth provided some relief.

Mr. Wu clasped his hands together and nodded his head toward Jay, as if to say "Thanks." His eyes sparkled conspiratorially, like the two of them were in on a joke.

"I was running all over looking for that thing," Jay said. "Our floor was all out of them."

The interpreter repeated this to Mr. Wu, who replied: "If it makes you feel better, I can use it for many different things." He pantomimed brushing his teeth with the sponge, and Jay giggled.

Mr. Wu had nearly died the week before. He was in his eighties and fractured his hip, and when he got to the emergency room, they discovered that he also had Covid pneumonia. Shortly after he went into septic shock, sending him to the ICU. His body was rail-thin, all knobby bones. He'd been put on an anticlotting medication because of Covid, but then he started bleeding internally, so the doctors stopped that medication. He kept telling them that he needed to get home to his grandchildren.

On the phone with Jay, his daughter had the same animated tone as her dad. "When do you think he might be discharged?"

"We can't say for sure yet," Jay told her. She was careful not to speak in absolutes about discharge dates. There was too much changing day by day, and it was better to overestimate recovery time than to set expectations too high.

Jay and her attending were checking in on another patient one afternoon when the attending got a phone call. She stepped aside for a moment, and then Jay heard: "Oh my God. He died?" The attending hung up the phone and took a deep breath, swallowing hard. "Mr. Wu died," she said to Jay, her voice catching.

The two ran upstairs to his room. There was a cluster of nurses and doctors around his bed. "He started agonal breathing," one of the nurses said. That was a type of open-mouthed, neck-extended gasp that meant a patient was nearing the end of his life. His blood pressure had dropped suddenly and his heart had stopped, possibly the result of a clot in his lungs. He was marked as DNR/DNI, so the team hadn't attempted resuscitation.

Without thinking, Jay pushed through the group until his body came into view. His mouth was open, and his freckled cheeks were flushed red.

Jay took Mr. Wu's hand. It was warm, a reminder that just minutes ago he might have clasped Jay's palm to thank her for the visit, his eyes twinkling. The line between living and dead was so thin in these thirty-eight-inch-wide beds. For a second Jay could almost see his chest rising and falling. It was all so sudden that it didn't seem real—he looked just like he always had. But it was only in her imagination. He was motionless.

Jay felt paralyzed, unable to muster words or movement. The attending physician turned to her. "Let's call his daughter," she said.

Mr. Wu's daughter picked up her phone right away. "Is this a good time to speak?" the attending asked. "Would it be okay if we share some serious news?" This was drilled into them in medical training; you wanted to make sure the soon-to-be-mourner wasn't driving.

The daughter was grocery shopping. "Yes," she said.

"Your father has died," the attending said. This was customary too. You had to use the word "dead" or "died," not a synonym or euphemism. It was brutal, but it was better than risking confusion. "I'm so sorry."

"Can I call you back?" the daughter said. Her voice tore open along the fault line of a sob. "I have to go."

The rules about hospital visitation were more lenient when a patient died, so Mr. Wu's daughter and his two grandchildren, who all lived with him, came to visit that afternoon. Jay led the group into the room. As soon as his body came into view, his daughter let out a scream. Jay watched as she rushed forward and grabbed his hand: "Daddy, I love you," she said. She began to shake, face wet with tears. "I love you."

The two children watched their mom sobbing. Their eyes widened as they looked on helplessly.

"I never got to say goodbye," the woman wept. "I'm so sorry, Daddy."

At these words, Jay's chest tightened. She realized that she might have been the last person to have a full conversation with Mr. Wu before he died. She thought of his gentle thank-yous, and the determination he put toward recovering so he could get home to his family. But this was the unspoken rule of the Covid hospital: no one got to say goodbye. It was all unfinished sentences and unjust departures.

"Goodbye, Daddy," the woman said softly, Mr. Wu's limp hand still in hers. "Have a good journey."

Jay's eyes were suddenly wet, so she turned away to compose herself. Then she felt a hand on her shoulder. It was Mr. Wu's daughter. They locked eyes. "Thank you so much for caring for my father," the woman said. Then she pulled Jay into a tight embrace.

There were so many people like Mr. Wu in hospitals all over the city—patients who died alone. Elderly people who coded the moment they arrived at the hospital, and faded as resuscitation efforts failed. Patients whose conditions deteriorated suddenly overnight, with their family members worrying over them in bedrooms just a few too many miles away. Fragile bodies slipping in and out of consciousness, while adult children were glued to their telephones, waiting for updates. Hospitals that had to say: *No visitors.* These were the rules, the cost of caution.

Many doctors worried about the long-term psychological effects of witnessing these solitary deaths. Amrapali Maitra, an internal medicine doctor in Boston, described it as a kind of moral injury: the feeling you've failed to prevent some event that contradicts your values. Dr. Maitra helped run wellness initiatives for the residents at her hospital. Phone call after phone call she heard the same message: *We can't take watching these lonely deaths.*

Periods of devastation tend to change our understandings of morality and mortality. In the 1340s, when the bubonic plague struck Europe and killed a third of the continent's population, death rituals disintegrated. Family members abandoned their sick loved ones, and even clergymen couldn't visit the dying. Their bodies were left for poor neighbors, who were summoned to dispose of the corpses for a meager sum. The city streets reeked of decomposing flesh.

It was during this period that people started circulating *ars moriendi*, handbooks that offered practical advice on dying. The little manuals, some illustrated, became an immensely popular genre. There was an emphasis on the social aspects of death. Friends and family were instructed to convene at the deathbed. It was the end of a hero's journey for the dying person (*moriens*), but the cast of characters surrounding them had a role to play too. They were expected not only to pray for the dying but also to contemplate their own mortality, in a rehearsal of sorts for their future deaths. Historian Philippe Ariès writes that in Europe, until the late nineteenth century, death was such a social event that even strangers might show up at the bedside once the death bell tolled and last rites were administered. The same still holds true of the postmortem weeks; funerals, wakes, and shivas are a chance not just to mourn but to gather.

While cities no longer have death bells to notify a community when one of its own dies, the sense that death should be a communal exercise persists. Not everyone fears death itself. Mortality is universal, one of the rare qualities shared across lines of class and geography. But the prospect of dying alone is almost universally frightening. People want to die surrounded by reminders of the love they built and lives they lived.

This is ever more true today, as evidenced by our visceral responses to reports of solitary deaths. In the year 2000, Japanese authorities went looking for a sixty-nine-year-old man who had stopped paying his rent and utilities bills. When they broke into his apartment, they found his skeleton near the kitchen, its flesh devoured by maggots and beetles. The man had been dead for three years. Though his apartment complex, or *danchi*, housed thousands of people, the man lived alone, so no one noticed as his corpse decayed and his home filled with the smell of rot. It was the first "lonely death" to capture the country's attention. Years later, a popular Japanese magazine reported on the phenomenon: "4,000 lonely deaths a week," it proclaimed on its cover. There was something about those two words when yoked together that tended to spark fear: *lonely deaths*. "The way we die is a mirror of the way we live," declared one of the dead man's neighbors.

So what were the Covid doctors to do when a patient coded and the family wasn't there? These patients weren't forgotten, like some of the lonely deaths in Japanese apartment complexes. But neither were they surrounded by a cast of supporting characters who could talk to them and help them reflect, or at the very least squeeze their hands. Before the pandemic, a patient's death in the hospital might include up to thirty-five family members packed into a room with musical instruments and prayers. During Covid, it was more likely to involve a nurse holding up an iPhone.

Tending to the Covid dying was logistically complex. As someone lay dying, their provider might be scrambling to look up family information. Sometimes the doctors had to leave voice mails—and had seconds to decide if they should give families their cell numbers or leave them to navigate byzantine hospital directories. Other times the doctors had to reach relatives abroad, desperately trying to dial Malaysia while the patient's final minutes ticked away. Then there were the cases of family members rushing from faraway coronavirus hot spot states who wanted to know if they had to quarantine before entering the hospital to visit a dying loved one.

"Questions came up that there were no policies for," said Hashem Zikry, an ER doctor in Queens. "Everyone thought there was a clear chain of command to decide. In reality, it was me sitting in a room thinking, 'Eh, am I comfortable with this?' "

One group of Covid doctors wrote an article for the *New England Journal of Medicine* expressing the horror of all the solitary deaths they had witnessed. "Dying alone and not being able to see your family members is something that runs to the direct antithesis of what we try to do in our roles," the lead author later said. "So this COVID pandemic shook us to our core in an unanticipated way."

Covid brought innumerable challenges to proper communication, so at some point one might suppose the doctors would become inured to the broken connections that came with practicing medicine at such a time. There were also the barriers formed by protective equipment, and the fears that made providers limit the time they spent at bedsides.

But there was something distinct about witnessing a patient pass without their family present. Sometimes, family members were able to make it to their loved ones in time, and the strict no-visitor rules were bent for those moments. All too often, though, they didn't. And as hearts and lungs failed, as families began to grieve, the doctors began their own process of mourning. Then they turned, right away, to the long list of others who needed their care.

Eighteen

GABRIELA, May 2020

After those hours of loan applications and then weeks of waiting came good news for Gabriela's family. Her mom had been approved for the government's Paycheck Protection Program. The money would cover continuing to pay her employees during the shutdown. Over the phone, Gabriela could hear the tension seeping out of her mom's voice. She was outfitting the salon with the safeguards it would need to reopen, ordering plexiglass and reading social media advice from other business owners about how they were handling their precautions.

"You're making your doctor daughter proud," Gabriela said with a laugh.

Meanwhile, the number of Covid cases at the hospital was dropping. New Yorkers were coming back into the emergency room with non-Covid ailments. It was a trickle at first, but then became a flood: abdominal pains, heart attacks, kidney problems. When Gabriela was treating a full floor of Covid patients, the work felt consequential and high-impact. She was needed; her fatigue didn't matter. But as the surge subsided, she was feeling the tug toward home and that pre-residency respite that her frontline weeks had disrupted. The exhaustion was catching up to her as she began her final week on the night shift.

Gabriela had learned that working the night shift often meant tracking the fluctuating heartbeat of patients she barely knew. The patients had their personal exchanges with the day team and were mostly asleep once Gabriela clocked in. Her role was to look out for any signs of impending crisis. Oxygen saturation levels dropping, blood pressure falling. With any shift in a patient's vital signs, she had to determine: Was it time to do something yet, to change someone's medication or call for a senior resident? She had to toe the thin line between a rapid response and an overreaction. The proper scale of alarm was especially tough for Gabriela to calibrate because she was new to adult medicine and to an intern's level of responsibility.

One of her patients that week was nearing the end of her life. The woman, Ms. Walker, was in her fifties and suffering both cirrhosis of the liver and a cancer in her nasopharyngeal cavity. She had lost so much weight that her collarbone and ribs jutted out, and each day her wispy brown hair grew oilier and more matted to her face. Her oxygen levels and mental faculties were in decline. She was combative and rejected all of the doctors' suggestions for her care, including sedatives. She refused to believe she was dying.

One night, Gabriela saw on the monitor that Ms. Walker's blood pressure had abruptly dropped low, to 80 over 40. She wondered whether she should alert a resident. When she went to check on her, the patient didn't look right. Often someone's *look* told you just as much as their vital signs. At 5:00 a.m. Ms. Walker's blood pressure plummeted again. Gabriela consulted her resident, and they decided to call a rapid response. She had never called one before, but she hoped it was warranted.

The ICU doctor came down to assess the patient. Gabriela tried to remain calm as she answered the doctor's questions about Ms. Walker's history: what medications she was on, what her course in the hospital had looked like. After a minute or so, she felt less jittery. She knew the answers the ICU doctor was pressing for. She could see that she'd been right to call the rapid response. They decided to put in orders for Ms. Walker to get fluids and albumin, a protein that retains fluid in the blood and maintains blood pressure.

It was nearly six in the morning, and Gabriela's muscles ached, but she felt a twinge of pride. Over the last few weeks, the unfamiliarity of various tasks on the floor had begun to fade—and now she'd made it through her first rapid response.

New York's mood began to improve by late May. Jazz bands serenaded city residents enjoying to go cocktails from local bars on the street. The mayor predicted that the city would be ready for the first phase of its reopening by June. Infection rates were dropping. And Gabriela's frontline assignment was winding to a close.

On her last night, she covered thirty-three patients. Gabriela remarked to a fellow intern that the team for one of her final shifts happened to be all minority doctors. Her attending physician was Hispanic, her pharmacist was Asian, and her co-intern was Black. Gabriela kept looking around and thinking, This is so cool. She'd never been on a full team of underrepresented doctors before. She also wished that the experience didn't feel like such an aberration.

Whenever Gabriela had a few spare minutes that night, she returned to sit with Ms. Walker, who despite her pain was still refusing any comfort measures. Gabriela was doing everything she could think of to keep the patient calm: gripping her hand as she grumbled, accepting her kicks and verbal abuse. It was her first time seeing someone die unmedicated, and it was both agonizing to watch and paralyzing. When Gabriela signed Ms. Walker out to a resident in the morning, she made it clear that the woman would likely die that day.

The residents were upbeat and congratulatory as Gabriela

packed her belongings. "Oh, you were an early graduate?" one of them asked her. "I had no idea!" That felt nice. But she also felt a heaviness as she headed for the exit. She passed residents in scrubs yawning as they started their day, and she thought about the warm hand of the woman who'd been nearing death that morning.

Ms. Walker's heart stopped right after Gabriela left, she learned in a text from her senior resident. Just like that, the frontline weeks were over.

Gabriela took a Covid swab at NYU, along with an antibody test, and began to pack suitcases for a trip back to Massachusetts. She and Jorge left as soon as the negative test result came in.

Her mom greeted them tearily. She showed off the Mother's Day present that Gabriela had sent earlier in May. It was a painting, oil on canvas, of a familiar face. Dark hair in a 1970s-style blowout, bold brows, lipsticked smile. Jewelry resplendent and smile both stately and coy, like a royal indulging her paparazzi. It was Grammy, their matriarch, in a painting commissioned from Gabriela's high school friend. It was hanging in the bedroom, over an armoire, where Gabriela's mom could see it every morning.

The family was more at ease now that the salon was back open and catering to a surge of clients with shaggy quarantine hair. Gabriela's mom divided the staff in two to cut capacity, and the receptionist took everybody's temperature at the door. There was no blow-drying—it was too hot for the stylists in their masks and PPE. The employees who had kids in virtual school got priority in choosing the timing of their shifts.

The relief in the return to the job was palpable. One of the salon workers had lost a grandma to Covid, and she told the others about saying goodbye over FaceTime while the doctors and nurses stroked her grandma's hair. All the clients were eager to hear about Gabriela's hospital work in New York City; one had brought in a newspaper clipping about NYU's early graduation from the *New York Times*.

One afternoon, Gabriela and her mom drove to get takeout

at a local burger stand. They pulled up outside and saw a swarm of bikers, a couple dozen, waiting to order. None of them were wearing masks.

"Oh my God." Gabriela was fuming. Her eyes scanned the group and landed on the young woman working behind the counter. Her mask was dutifully pulled over her face while she served these men, who were laughing and calling to one another. These mass vectors, Gabriela thought.

"I need to go," Gabriela said. "I'm about to lose my mind." She turned away and rushed to her mom's car. Locked inside, she took a deep breath and her eyes started to tear.

Gabriela was so tired. Tired of having to do the high-risk, frightening work that she was called to do, only to see groups of people who couldn't be bothered to put on their masks. Tired of seeing apathy, in high definition, even while the hospitals were still overflowing with people who might never wake up.

"I thought you were going to have your Norma Rae moment," her mom said later, ribbing her gently.

Maybe these unmasked bikers didn't think it was a big deal, Gabriela thought. Maybe they told themselves that their actions didn't have real effects, so they could roll their eyes and adopt a cavalier air. But there were people who would carry the cost of that, like the patients Gabriela saw at NYU Tisch. There was always someone who had to bear the burden of other people's choices, even the minor ones.

Sometimes Gabriela's mind returned to the wry smile on the face of the white attending physician who called her *mi amiga*. She thought about the voice inside her own head that said she couldn't lash out. For some people these were just little jokes, or little trips outside without caution. The toll they took could seem invisible. But others didn't have the luxury of negligence. Gabriela had seen the cost of carelessness throughout her frontline weeks, and especially in the faces of her patients who passed and the family they left behind.

At 160,000 square feet, the NYU Hassenfeld Children's Hospital is a glassy behemoth, a fit-for-brochures facility that peers out over the East River, with a thirty-foot-tall Dalmatian sculpture standing guard at its entrance. At night the building lights up like a Christmas tree; during the day it sucks in and spits out a stream of interns, residents, attending physicians, nurses, and patients. Gabriela would spend some of her internship weeks in that tower. And it was in the courtyard outside Hassenfeld, days before the start of her residency, that Gabriela gathered with two hundred other NYU doctors for a vigil.

It was less than a month since George Floyd had been killed in Minneapolis when a white police officer kneeled on his neck for eight minutes and forty-six seconds until the forty-six-year-old Black man died of asphyxiation. A bystander's camera had caught it all. In his last minutes, Mr. Floyd had begged for his life. "Mama," he was heard to say. "I'm through." It was a primal cry that fueled a surge of participation in the Black Lives Matter movement, possibly the largest movement in American history. In the early summer of 2020, more than 15 million people across the nation marched.

In New York, Gabriela gathered with her NYU coworkers to show their solidarity with those millions. The pediatric interns filed onto the pavilion outside Hassenfeld. Dozens of other residents and faculty members packed in around them, some in their scrubs and others in white coats and slacks.

Several doctors held up posters with statistics on racial health disparities. One of these signs, which caught Gabriela's eye, read "Black babies are more than two times as likely to die as white babies in the neonatal ICU." Doctors were moving to connect the protests against police brutality with the medical field's long history of racial bias and abuse. Black patients had worse health outcomes. They were less likely to have generational wealth, which limited their ability to access preventive care, nutritious food, and even stable housing. All of this meant the average life expectancy of an African American person in the United States was significantly lower than the average life expectancy for white people. New York City's Health Department had recently released a statement recognizing that racism itself was a "public health crisis."

Between February 29 and June 1, New York City had seen more than 200,000 people diagnosed with Covid, 50,000 people hospitalized, and more than 18,500 who died. The more poor a neighborhood was, the higher its hospitalization and death rates. The virus hit hardest in the city's Black and Hispanic communities: 699 of every 100,000 Black New Yorkers were hospitalized, and 658 of every 100,000 Hispanic people. For white people, that number was 314 per 100,000. Many of those who got infected or died were essential workers—subway operators, food delivery people—unable to isolate while the well-to-do drew indoors.

It was hard for Gabriela to hear the speeches from across the pavilion, especially with the roar of city traffic in the background. The organizers didn't seem to have expected such a large turnout, and there were no microphones. But then someone stood and called for eight minutes and forty-six seconds of silence. The stillness that ensued was deafening. Everyone lowered to their knees. Gabriela looked around at the masked faces and kneeling figures surrounding her. She felt a swelling sensation. These were her people. They had rushed into emergency units as the city fell sick; they had stepped away from their work to take a knee in the face of a different kind of crisis.

Gabriela had learned that her pediatrics intern class was one of the most racially diverse in the program's history: five of the twenty spoke Spanish, four were Black. It was again somewhat dispiriting that this should feel notable, but it did. Gabriela had decided to join the Pediatrics Diversity and Inclusion Committee, a group of doctors at all levels of seniority in NYU's pediatrics department who were focused on recruiting and retaining doctors of color, and addressing inequities for both providers of color and their patients. They had held a welcome meeting over video conference on June 10. There were subcommittees on recruitment, mentorship, education, and outreach. Gabriela and her friend were made the point people for the recruitment branch, which would mean planning virtual social hours for Black and Hispanic residency candidates. The group was also working on developing a system for tackling microaggressions in the department—which would start by educating doctors on what microaggressions in the hospital actually looked like. As Gabriela prepared to start her professional pediatrics training, she'd been thinking about how a pediatrician's role was to get involved in every aspect of a young patient's life, including family history and social ties. With this level of intimacy, it was all the more important to ensure Black and Hispanic children were receiving equal care and attention.

Gabriela began to wonder what her grandmother would have had to say had she been alive through the past months of pandemic and protests. Grammy would have had some choice words for the country's president, that much Gabriela knew. Grammy used to tell the family about the prejudiced comments she received because of her Middle Eastern roots. At the hospital where she worked, there were other nurses who had called her "darkie." She would have been proud of her granddaughter now, of how Gabriela chose to show up.

Suddenly the eight minutes and forty-six seconds of silence broke. The nurses and doctors began to rise to their feet again. Back to the hospital floors, and to a different sort of front line.

Nineteen

JAY, May 2020

As the end of Jay's assignment approached, she kept wrestling with ways that she could say goodbye to Manny. This was the toughest aspect of the hospital: patients drew as close as family, and then you had to cut off the tie. You were reminded, all of a sudden, that the relationship wasn't permanent. It wasn't made to last longer than the toughest few weeks of someone's life. In fact, its severing was often the best news the patient could receive. It usually meant that they could go home.

Jay didn't want to become one more figure in Manny's life who disappeared. She had hoped that by the time she was leaving, Manny, too, would have left the hospital. She wanted to see him settled into one of the city's group homes. But as her final week rolled around, Alicia was still filing applications and setting up Zoom interviews. There was a home in Far Rockaway that seemed promising, but they hadn't yet received a definite answer.

On Jay's last day, when she went to find Manny, he was sitting in the hallway with the PCA Paul, whom he called Papi. They waved as Jay approached. Manny was wearing the crisp collared shirt one of the nurses had given him.

Not yet ready to formulate her last words, Jay turned to Paul and asked how he was doing.

Paul said he was worried about Manny, who kept gesturing toward his neck as if to signify a sore throat. Jay told him some of her patients with discomfort in their throats actually had mouth ulcers, which she treated with what was called "magic mouthwash," a blend of hydrocortisone, nystatin powder, and diphenhydramine HCL syrup. She liked talking shop like this, sharing medical knowledge as she gained more confidence in her own instincts.

But after twenty minutes, Jay knew she was prolonging the inevitable. She was glad to be able to say her farewell when Manny was sitting with Paul. She looked at Manny, who was happily taking in the scene of his favorite staff members chatting.

"Manny, it's my last day," Jay said to him in Spanish. He cocked his head to one side, still smiling at her. "I'm going to miss you," she said. "*Cuidate*, take care of yourself."

He reached out his hand, as he always did when she was saying goodbye to him. She grabbed it, and they interlaced their fingers. Then she gave his palm a squeeze. "It was a blessing to know you," Jay said. "Good luck, my friend."

At some point that day, Jay recalled the prayer that her mom had sent out to friends in an email just before she started work in the Covid wards: May the angels that helped clear her path thus far be forever at her side. It was a note typed out with so much worry. If only her mom had known that there were angels waiting in the Covid wards the whole time.

Jay's residency in pediatrics and internal medicine was in Springfield, Massachusetts, near the hospital that treated her brother's ski injury when she was ten years old. In the days after her work finished, she packed all her belongings into boxes. She sifted through med school papers and looked for knickknacks she could discard. Then she shoved the rest in her car and headed for Springfield.

When she pulled into the driveway of the house where she would live, she was surprised by a familiar sight outside: her mom. Kate had driven the same mileage up as a surprise.

"How could I not help you move into your new home?"

Jay got out of the car. She wanted to hug her mom, but instead they faced each other from six feet away. Jay thought about all that had changed in their months apart. All the people she cared for. All the families she saw warped by Covid, finding language to discuss the impossible—not unlike her own.

Jay could tell that her mom had just needed the satisfaction of coming to see her in person and reporting back to the rest of the family that she was safe, healthy, and settled in her new house. Within an hour, Kate was back on the road to New York.

Jay's new house was big and white with navy shutters, a sunlit space that made her city apartment look like a bunker. Inside, her new roommates, whom she'd met briefly on a quick visit to Springfield earlier that spring, had left a trail of gifts: a bottle of sweet white wine and a note written in colorful marker that read: "Thanks for helping fight Covid!" Their puppy, Poe, ran in frenetic laps around the kitchen.

As Jay started her fourteen-day quarantine, the weeks in the Covid wards began to feel distant—as hard to recall as those early days before her frontline work, when thermometers were in short supply and people joked about making their own hand sanitizer. It felt separate from her new life in a hazy, dreamlike way.

Sometimes the residents she'd worked with sent her text updates on Manny. One day she learned the good news: He'd been admitted to the group home in Far Rockaway. On the day he left, his hundredth in the hospital, dozens of doctors and nurses came to applaud him. One of the physicians brought him a guitar, and another came by with tambourines to put on an impromptu concert. The hospital halls were suddenly jumping with the sounds of "La Bamba."

Jay was relieved when she started residency training and her days resumed their structure, though she kept having to remind herself to pay attention during the long introductory lessons, including on donning and doffing PPE. The seven other interns in Jay's program—the first-ever all-female cohort, they were told—were jittery about the transition from home quarantine to the hospital. One by one as they heard Jay's story, they leaned over with reverence: "You're coming from *New York*?" each one asked. "You were in the *Covid wards*?"

Jay shrugged them off. "It actually wasn't so bad," she kept saying, smiling when she remembered her own shaky hands on the first day at the hospital earlier that spring.

The residency director pulled her aside during training one morning. "We're putting you on outpatient at the beginning," she told Jay. That meant slightly looser hours, staffing the day clinic instead of the inpatient hospital floors. "We don't want you to burn out coming straight from New York."

On Jay's first day of residency, she woke up at 5:30 a.m. with an eerie sense of calm. Her first patient was a five-year-old girl whose hair was tied back in a red bow. Her mom only knew Spanish, but the girl spoke a hybrid of the two languages, English phrases punctuated by exclamations that her mom could understand. At first the girl fixed her eyes on Jay in a shy silence, but once she saw Jay's smile, she opened up.

The girl's mom told Jay that her daughter had a sharp pain that alternated between her ankles and knees. It troubled her especially when she walked. Jay asked her to do a lap around the office. Then she launched a series of questions aimed at process of elimination: Could she have a torn ligament? Or a fractured bone?

"I actually think it's just growing pains," Jay told the mom finally, in Spanish. She looked at the girl. "Do you know what growing pains are?"

The girl shook her head.

Jay explained that growing pains resulted from the natural process a body undergoes as it adjusts to someone growing up. What she could've added was that they often come later in life too. People are always in a process of maturing, settling into new environments with all the attendant aches. Stretching past the limitations of who they were before, or who their parents thought they could be. Jay would know.

BEN, May 2020

In mid-May Ben had a patient who he really thought could pull through. Maybe. Mr. Joseph had shown up in the emergency room with Covid on top of chronic kidney disease. He would need dialysis to remove the waste building up in his system. He wasn't as elderly as many of the team's other patients—he was in his sixties—so with quick action from the team, Ben thought he could survive, unlike so many of the other patients they had seen on the telemetry floor.

Ben came by the room to introduce himself and found Mr. Joseph, a large white man, lying down and staring stone-faced at the wall. "Nice to meet you," he said. "I'm Ben, I'll be taking care of you."

Ben had noticed there were no emergency contacts in the patient's file. "Who would you like us to call if something happens to you?" he asked.

"No one," Mr. Joseph said tersely.

When Ben brought up his worsening kidney function, Mr. Joseph quickly declared his opposition to dialysis. "My nephrologist has been trying to get me to start it for years," he huffed.

He told Ben that he didn't want to spend his life at the mercy of invasive machines.

This was the kind of patient who prompted a response that wasn't clear-cut. He needed immediate action for his kidney disease, but he didn't seem open to it. He had no family members calling to ask how he was feeling and offer solace from afar. Ben worried that Mr. Joseph seemed to have lost his will to live.

Ben had a standard explanation that he told patients who seemed upset or confused about the idea of dialysis: "The function of your kidneys is to filter the blood. Dialysis is a machine that does it for you. With hemodialysis, you go three times a week to a center where your blood is taken out of your body and passed through a filter. With peritoneal dialysis, you can do dialysis at home, by pumping fluid into your belly."

Mr. Joseph's stare was flat. He listened in listless silence, as if to say: *Are we done here?*

After a few minutes, Ben had to leave to check on other patients. But as he retreated down the hallway, he wondered whether there was anything else he could have said to make Mr. Joseph more comfortable with the treatments he needed.

A few hours later, Ben was at his computer in the workroom when the renal doctor called for him with news, after having gone to meet with Mr. Joseph.

"He's agreed to dialysis," the doctor said.

Mr. Joseph still had a chance. It wouldn't be an easy road. He would have to undergo a procedure to have a temporary catheter placed in his neck for urgent dialysis. Until he underwent dialysis, his kidneys would continue failing to filter his blood and he would be at high risk for an arrhythmia—a heartbeat that goes awry—which could be fatal.

Still, Mr. Joseph had overcome his hesitation and fears about a potentially lifesaving treatment. Ben felt a momentary lightness when he left work that evening.

He walked back to the dorm, as he always did. During these

five-minute commutes, the city's purple twilight seemed to tease the gray of the hospital shifts. Inside, Ben's patients were crashing and their families were grieving. Outside, the spring was softening into pastels. Businesses that had closed at the peak of the pandemic were reopening, this time with signs that read "No mask, no service." There had to be some German word for a city turning luminous in the midst of disaster.

The next morning Ben hurried through the lobby and temperature check. He was eager to get to sign-out with the night team, check in with Mr. Joseph, and discuss preparations for dialysis. He still felt buoyed by Mr. Joseph's about-face on the kidney replacement therapy. He had seen so many patients undergo lifesaving interventions that had only the slimmest prospect of working. This seemed like a slightly more promising effort.

Ben could tell within moments of gathering with the night team that something had gone wrong.

"We had a patient pass overnight," one of the residents told Ben. "We didn't have any emergency contacts on file."

Mr. Joseph had gone into cardiac arrest unexpectedly. The night team had found one old family phone number in his records, but when they tried it, no one picked up.

As Ben listened to the resident, he felt grief radiating through his body. It was the realization that he might have been one of the last people to speak with Mr. Joseph, coupled with the knowledge that there was no one else to mourn the man's loss. Ben had left work the previous day thinking Mr. Joseph's fight would continue, but sickness had won out in the end. The team debriefed rapidly; they all knew there wasn't anything they could have done differently.

Hours later the hospital received a call from Mr. Joseph's brother, who had heard that his sibling was sick.

"He died last night," Ben said. "We didn't have any family information on file."

"Wow," the voice on the other end of the line said, with distant remorse. "We didn't even know he was sick."

"This was a hell of a way to start your medical career," one of Ben's residents told him.

Ben nodded. His team had seen twelve deaths; six of them were patients Ben had followed.

He was relieved to have a few weeks of quarantining before he was set to start residency in July. He had a virtual emergency medicine course to complete in preparation, and after that he'd have some time to hike and rest. For his residency, he would be moving from internal medicine back to emergency medicine. He looked forward to it. The truth was, he missed everything about emergency care. He missed the fast pace: clock in, get your patients, assess their needs, get the follow-up imaging or blood work, reassess, and repeat. He missed being the first doctor a patient spoke with when they walked in the hospital doors.

He missed the medical bread and butter. Oh man, he missed the bread and butter. Abdominal pains, chest pains, syncope, asthma, COPD, cellulitis, gastrointestinal bleeding. Dehydration! What he would give for a simple case of dehydration, goddammit: an IV with some fluids and a recovered patient on their way home.

He missed visiting his patients without the plastic layer of a shield and an N95. Grasping a hand or touching a shoulder. Looking around the hospital without the virus lens—that gaze that turned everyone into a vector, a walking and coughing mass of infectious particles.

Still, Ben told himself that his weeks of Covid care had left him better prepared for anything the next phase brought. Especially tricky conversations with family. By now, he knew how to talk about death. That was one thing the Covid floors had taught him.

The night before his last shift on telemetry, Ben decided to prepare a parting gift for his team. His go-to gifts were cocktail mixes.

For the telemetry team he made grapefruit soda: a blend of grapefruit juice, water, agave syrup, and lime juice, with a pinch of salt. He had a stock of mini champagne bottles that he used for these sorts of presents. Each held 6.3 ounces. When Ben gave these out he liked to tell his coworkers you could drink some during the day, then save the rest to mix at home with tequila for a highball.

It was 84 degrees outside on his last day. The glass of the hospital building glinted in the early-morning light. Ben's bag jangled as he moved up the steep ramp outside, a chorus of mini bottles singing at his side as if he were a retro neighborhood milkman.

At the morning meeting, Ben's teammates gratefully accepted their bottles of grapefruit soda.

"Hell of a job," they told him.

"Not an easy month," one said.

It felt peculiar finishing his stint this way: warm day, coworkers sharing words of praise, everyone tipping back their bottles of syrupy soda. The day had the quality of a cordial, languid and saccharine. He discharged one of his patients, a woman who had come in with a central line infection. By 2:00 p.m., he had rounded and finished updating all his notes.

"Why don't you head home?" his attending told him. "Go enjoy your afternoon."

Ben shrugged off the layers of protective gear. He stripped off his dirty scrubs. He walked out of the hospital into the thick afternoon air, the glassy fortress receding behind him as he headed home.

In times of disaster, it's hard to resist putting our new emotions into old frames. "One cannot think without metaphors," Susan Sontag writes. War narratives offer themselves most readily. In the nineteenth century, bacteria were labeled "agents of disease."

For decades cancer inspired military language. "Cancer cells do not simply multiply; they are 'invasive,' " Sontag notes. "Cancer cells 'colonize' from the original tumor to far sites in the body, first setting up tiny outposts."

The AIDS epidemic, too, became a "battle" to "win" in the popular imagination. "The survival of the nation, of civilized society, of the world itself is said to be at stake," Sontag wrote. The illness—or enemy—is invariably imagined as a foreign intruder. One illustrative case, according to Sontag, is the syphilis epidemic that swept Europe in the fifteenth century. It was the French pox to the British, *morbus Germanicus* to the French, the Naples sickness to the residents of Florence, and the Chinese illness to the Japanese. The metaphor of war functions best when the foe is the other, encroaching on native land. Leaders deploy it to conjure a sense of unity. It's us against them. Good versus evil, right against wrong. The deaths suffered become almost inevitabilities, the collateral damage of a righteous fight.

In some ways, the Covid Coalition's experiences weren't unlike battle wounds. They were young people called to serve their communities. They witnessed loss. They carried guilt. They felt the sweet togetherness that comes from being bound up with the people around you in a fearsome, gallant fight. They tried to unravel the truths and the sorrow spooling out over just a few fleeting frontline weeks.

They walked out with scars that will shape their careers, lessons that some of their seniors absorbed only over the course of decades. They gained a new view of heroism: when their role wasn't to save a patient, but to help them determine how they wanted to die. Humility: when the most powerful force they could bring into a hospital room wasn't their technical skill but their desire to connect. And loss: when there was nothing to be done to stem the tides of a patient's grief, or a coworker's, or their own. More than seventeen hundred health care workers died in the United States in the early months of the pandemic.

But their work was never supposed to be like warfare. Armies don't ask for sacrifice so much as demand it. When soldiers enlist, they are typically prepared for offensive strikes, to do harm and put themselves in harm's way. That is not meant to be a doctor's role. Doctors mitigate danger. Their end goal is safety.

The soldiers who make it home from battle sometimes have to contend with moral injury. They wrestle with the senselessness of the suffering they witnessed. To a doctor, little is senseless. They can trace the molecular shifts that lead to a body's collapse. They know every joint and organ in their terrain. They often have ample information. Sometimes they know enough to realize there isn't any further intervention they can try. There is no General Custer's Last Stand, no pride in a futile defense like the Alamo's. Just quiet acceptance of bleak odds.

And for the patient, this battle-cry language can bring its own pain. Framing sickness as war can mean the dying have somehow failed, or lost something that might have been won. Linguist Elena Semino, who studied the way metaphor is used to discuss cancer, found that aggressive language can make patients feel like they've done something wrong in being unable to stop the disease's spread.

The Covid recruits weren't navigating the soldier's territory of wrong and right. They were facing thorny questions with no certain answers, and party to conversations about impossible choices. Circumstances that rejected all their training and clinical reasoning. War is often framed in terms of moral absolutes—especially "the last good war" fought by those Allied forces abroad. But so much of Covid care unfolded in a gray area. In some respects, it was a time marked by sentiments thicker than rosy, duty-bound patriotism. It meant tangling with mortality, with the limits of medical power, with physical and emotional bonds more palliative than drugs and machines.

Amid the pandemic, it was only natural we reached for metaphors. Familiar words help us process alien moments. The coronavirus was novel, and our responses were improvised; telling war stories, meanwhile, is an exercise we know well. But sometimes metaphor can be a mask. Underneath words like *coalition* and *battle* lies the reality of six young people just doing their jobs. They took an oath. They took deep breaths. And each morning they put on their scrubs and went to work.

Twenty

IRIS, June 2020

The body trucks outside Montefiore dwindled in late May. For all the drama of the early weeks, Iris's last days on the wards were quiet. There were fewer Covid cases. She couldn't help feeling guilty when she heard the 7:00 p.m. applause outside. At this point, the Covid census was so low that all the tributes to "health care heroes" felt almost comical. Iris referred to the dinners donated from local restaurants as her "guilt salads."

Most of Iris's new patients were suffering from ailments they'd tried to ignore while the hospitals seemed unsafe. There was a woman with excruciating abdominal pain and constipation. There was a man with cellulitis who waited so long to come to the emergency room that he turned up with his body in sepsis, his organs overwhelmed. "This isn't your typical cellulitis case," Iris's attending told her.

Her last day was uneventful. At Montefiore, residents typically switched teams every two weeks, so they were used to turnover. Iris huddled with her team in the morning, as usual. "We'll see you back here in a few weeks," the residents told her. By July she would begin her internal medicine residency there.

That afternoon her attending sent her home early, and Iris walked out of the hospital into a blinding spring day. She stopped by a local deli that had been sending free food for the health care workers so she could thank the staff there in person.

Then Iris and Benjamin took a celebratory drive to Flushing, Queens. Crossing the Long Island Sound, they took in the city skyline piercing the clouds, that view both familiar and cinematic. They picked up Kung Fu tea, milky black with gelatinous tapioca balls, and bought bok choy that was crunchier and more fresh than any you could find in the Bronx or Manhattan. Iris realized how much she had missed their old normal—leaving their apartment to shop for produce like they used to do on Sunday afternoons. Flushing's food markets were slowly reopening, though it was eerie seeing its thoroughfares so quiet. Main Street was usually a destination unto itself, lined with shoppers gossiping and shepherding along their small kids. Now it was dotted with just a handful of heads-down pedestrians.

But in the days after Iris finished work, the city sprang out of its silent hibernation. Suddenly the Bronx was not just awake but quaking. The avenues filled with fireworks and footsteps. At all hours, Iris could hear the chants: "No justice, no peace."

At first, along with her pride in this movement, she also felt nervous. She could see all the points of possible transmission at these demonstrations: people marching inches apart, shouts morphing into spit. It was hard not to imagine a spike in cases weeks later, leaving Montefiore once again overwhelmed. The second wave could come in Iris's first days of residency.

But mostly that stress was subsumed by hope. People were being pulled out to the streets by the same thread that had drawn Iris into the wards: a sense of obligation. Iris kept thinking about the Black communities hit hardest by the pandemic, her patients fighting for oxygen as their breathing tubes came out. There was enough suffering these last months without adding human-caused tragedies to the mix.

Iris was also beginning to map her own place in the country's racial reckoning. She knew that her family's story in this country was one of struggle. Her parents had uprooted their lives from Shanghai and come to a foreign country so that they could have more opportunities, both educational and economic. Their community battled stereotypes and bias, ever more so in the months of the pandemic. In the evenings Iris had started reading *Minor Feelings*, in which poet Cathy Park Hong wrote of her Asian American identity:

> The indignity of being Asian in this country has been underreported. We have been cowed by the lie that we have it good. We keep our heads down and work hard, believing that our diligence will reward us with our dignity, but our diligence will only make us disappear. By not speaking up, we perpetuate the myth that our shame is caused by our repressive culture and the country we fled, whereas America has given us nothing but opportunity. The lie that Asians have it good is so insidious that even now as I write, I'm shadowed by doubt that I didn't have it bad compared to others.

Park Hong's message resonated with Iris, who felt that Asian Americans had gotten lost in America's racial landscape. Pigeonholed as the "model minority," her community was pitted against other racial minorities. There was little space or opportunity to process the traumas inherent in Asian American experiences like the thousands of Chinese people who were killed while laying tracks for the transcontinental railroad. For every two miles of track, three Chinese laborers were killed by dynamite, by snowstorms. This was a trauma that came generations before Iris, but it was still part of her people's history in the United States, too often unstudied and unacknowledged. Iris started to feel that her community was the country's racial wild card, thrown away if it wasn't needed and exploited when convenient to benefit those in power. She hadn't experienced consistent racial hardship, nor racial privilege; she occupied a liminal space between.

At the same time she knew there were certain invisible forces that had allowed her to succeed as the daughter of Asian immigrants, especially her own community's ample representation in the medical field. She had worked with many Asian American residents and attending physicians. She had seen Chinese-speaking patients come into Montefiore and benefit from the plethora of residents on call who happened to speak Mandarin. Black and Hispanic doctors, meanwhile, were fewer and further between. They faced discrimination that Iris hadn't encountered in her work. During medical school Iris had read sociologist Robin DiAngelo's *White Fragility.* One of her white classmates had assumed Iris was reading it as a person of color, and she had to explain that actually she'd been drawn to it less as someone racially oppressed than as a potential oppressor understanding her privilege.

"Racial trauma is not a competitive sport," Park Hong wrote. This was true too. Iris had layers of privilege and struggle all at once. She understood that this was the dance she was bound to do, making sense of the ways that she marginalized others and simultaneously was marginalized.

Iris also kept thinking about the Black and Hispanic patients she saw during her med student rotations in the Bronx who'd been failed by the city and the health care system. Frequently they lacked access to affordable housing, healthy food, and insurance. As a provider, Iris was hamstrung. She couldn't buy them the healthy food they needed to help manage their diabetes or high blood pressure. But now as the protests against police brutality mounted, her course of action was clear. She could march.

On a Saturday morning, with crowds protesting all around the city, Iris and Benjamin decided to rally in their neighborhood. They joined the throngs packing Pelham Bay Park. People carried signs that read "I Stand with Black Lives Matter" and "Racism Is the Virus." A trail of five cop cars followed behind.

Iris felt a comforting degree of anonymity; unlike her work in the hospital, here she was contributing just by showing up.

Iris and Benjamin walked and chanted, taking in the faces around them. After forty-five minutes, the skies opened into a sun shower. Iris exchanged looks of half panic and half delight with the protesters around her as they were pelted with droplets, hair matted to faces and clothing clinging to skin. They moved forward, continuing their chants: "What do we want? Justice! When do we want it? Now!"

Iris had forgotten what it was like to be in a mass of people, the electricity of it. The last time she'd been in a group this large was back in March, at the airport, the day the reality of the pandemic had set in. That was a crowd animated by a different kind of charge. That group pulsed with panic, while this demonstration thrummed with hope.

ELANA, June 2020

Elana's dad spoke his first word shortly after the breathing tube came out of his throat. The doctor called in a rush of excitement. "That's good," Elana said, but she was only cautiously optimistic. During her time away from the hospital helping to care for family, Elana had been studying her dad's condition. The first word her dad spoke was an expletive, and she had read that the mind codes and stores profanity differently than other language, deeper in the brain, which means it can be retrieved more reflexively. People with traumatic brain injuries can lose vast parts of their vocabulary but retain the capacity to curse. So she wasn't yet ready to celebrate the hospital report.

But two days later a more encouraging call came. The neurologist had assessed her father's mental function, and said he had the potential to make a full recovery. This time Elana's eyes welled up. She could remember, with a clarity both painful and poignant, the moments in the hospital when she thought her father would die. She could remember reviewing the facts of his case with an attempt at clinical remove: forty-five minutes of CPR, no oxygen to the brain. His chances of survival were so low that there was very little research available on the likelihood of someone in his condition regaining brain function afterward. That he would retain his old personality seemed improbable—but she knew that this doctor on the other end of the hospital line wouldn't offer false hope.

The next few days were a stretch of endless nerves. All her family's household exchanges were taut. There were no more medical decisions for Elana to consult on. Now she had to wait. The doctors called with news that was mostly banal: a bout of agitation, a wiggle of his toes. A urinary tract infection one afternoon. There was little for the team to do, beyond watching his body and breathing grow stronger. Because he was now on the medical floor, out of the ICU, there were also fewer nurses around to FaceTime Elana and check on her dad throughout the day.

This was Elana's least favorite part of medicine, the waiting game. It was nail-biting on rotations, but excruciating now. It was a reminder that the power of medicine was only partial. The doctors could play God, but only up to a point. Even with infectious diseases like Covid, much of what the providers could do was manage emergencies and then assist while the body mounted its defense.

Having Akiva at home helped. Anytime there was tension in the household he appeared, magically, with a video game or joke. He took Elana's brother out for driving practice. He had an unparalleled capacity to create a sense of normalcy out of chaos; Elana liked to say he would be everybody's first-round draft pick for an apocalypse bunker. Meanwhile, Elana went on walks with her sister Daniela, looking for huckleberries, and learned new video games from her younger brother. She played bad cop when he asked to see friends or go to a neighborhood birthday party.

"Social distancing," she reminded him, ever more important as the possibility of their dad's discharge drew closer.

On FaceTime, it was clear to Elana that her dad's brain was still clouded. He often got confused about the year or mixed up her siblings. Once Elana's mom asked him how many children he had. "Five," he responded, and then counted them out on his fingers, including Akiva.

"Akiva isn't your son," Elana's mom said. "He's your son-in-law. He's married to Elana."

Her dad paused, processing this news. "They can't be married," he said incredulously. "They're siblings!"

He sometimes forgot how old Elana was, thinking she was five, and she had to remind him that she had graduated from medical school.

"You graduated? You're a doctor?" Each time he heard this, he grew more amazed, his eyes moving from Elana's face to the white-coated figures around him in a dance of mystified word association.

They had only two hours' notice when they learned he was finally coming home. Elana and her siblings scrubbed down the house, knowing how much calmer their dad felt when the home was tidy. Akiva drove to the grocery store and bought enough food to feed a small army.

They made tacos for his first dinner. Every bite was a reminder of the mundane bliss of eating tortilla shells with their dad again. Elana kept watching him as he forked up his food. Every micro moment felt sacred and fragile.

After they ate, Elana measured out her dad's drugs. She studied the paper with instructions for his rehab exercises. She would now play a doctor-like role—reminding him to do physical and speech therapy and take his medications. Wishing that she'd had training as a nurse, she kept calling her grandma for all the questions that no one thinks to teach you in medical school: Where can I safely store the drugs? Does the temperature it's stored at actually need to be in this exact range? If his blood sugar is high, should I give him another two units of insulin?

Day by day, he grew more aware of his surroundings. This made him more frustrated with his reliance on everyone around him—his inability to walk much, or bike like he used to. He had lost twenty pounds in the ICU, mostly from deterioration of muscle mass. He also had acid reflux; the dividing line between his esophagus and stomach had loosened during his time in the hospital. Elana wasn't used to seeing her dad sapped of all his energy. There were days when he got overwhelmed with fear, thinking of the hour he'd hovered near death.

They fought sometimes when Elana reminded him to do his exercises or told him not to have ice cream after dinner. It was disconcerting to assume the authority of a medical figure when talking to her own dad. She was used to coming to him with her questions and watching his face adopt its posture of certainty and wisdom. Now their roles were reversed. But he always relented eventually, retreating to a corner of the living room to stretch and breathe as the therapists had suggested.

After all that Elana's dad had been through, she and Akiva decided they should call a lawyer to formalize their own living wills. There were studies that showed that if you didn't tell your family members exactly what kind of end-of-life care you wanted, they were no better at guessing than a stranger. Elana chose Akiva as her health care proxy, and as a backup she selected her mentor Sharon, who she knew understood all the medical information that might color Elana's preferences for end-of-life care.

Both Elana and Akiva decided they wanted to be full code: in the event of disaster, do everything. When they talked it out, though, Elana determined that if she ended up on a ventilator for more than six months or a year, and it didn't look like she'd make a full neurological recovery, she'd want to cut off life-sustaining treatments. She didn't want to live like a vegetable. She'd seen enough things in the hospital to know that about herself.

Elana's Orthodox community was strict about the activities that men and women don't do together, like swim or sports or prayer. At the various synagogues she went to, growing up, the dividing line between the sexes was sacrosanct.

For the teenagers who didn't fully understand the boundary, social pressure served as its own magnetic force, drawing everyone to their places and repelling those who might have been tempted to cross a line. But the youngest children were too young to comprehend. They darted back and forth across the dividing line, flinging themselves around their dads' legs or into their mothers' laps.

Elana was once one of those kids. When she was little, too little to care about rules and judgmental eyes, her favorite thing to do was to run onto the men's side of the sanctuary and tuck herself under her father's *tallit*, his prayer shawl. Lulled by the dips and crescendos of the voices around her, she would look up at the cloth of translucent white above her head, and beyond that the jagged cliff of her father's chin.

As she listened to him pray, he would rest a hand on the top of her head, tousling her hair. Sometimes he would peek down and give her a furtive smile. Those were the moments she felt closest to her dad, witnessing his devotion, enveloped inside of it.

In the endless days of his hospital stay, Elana didn't pray much. It wasn't that some thought of a divine watchful eye didn't cross her mind; it did. But Elana had never believed in the type of prayer dredged up in moments of personal crisis. It seemed transactional. Like an exchange of fealty and flattery for favoritism. It didn't fit into Elana's theology, her vision of a God who didn't have petty preferences. God had to have bigger things than her family in mind, she figured.

Elana had learned this from her dad. "I don't believe in a foul-shot God," he used to tell her, meaning a deity who only appeared when a game was on the line. So she preferred to place her faith in the hands of the hospital staff. That was why she was training to be a doctor, after all; because she believed in medicine more than miracles.

But when her father returned home, he was set on resuming his daily prayer ritual, and he needed someone in the family to help him stand on his feet. Elana volunteered. His first morning at home, he asked Elana to go find his phylacteries, or tefillin, the small leather boxes containing parchment strips inscribed with scripture that are bound to the arms with leather straps. She watched him wrap them around his arms, and then she steadied him as he began to sing.

"Blessed are you Adonai, our God, King of the Universe who gave the heart understanding to distinguish between day and night," he chanted. He rocked back and forth, and she kept her arm around him, holding him up. Her dad's voice shook as he retraced those familiar words, words that he had said every day for his entire life until the frightening past few weeks.

He thanked God for the gift of life. He thanked God for the gift of his breath, and his body. Elana's eyes welled up. It was the first time she had seen him pray close-up since she was five years old.

In that moment, Elana felt lucky to be of use. Her Covid spring wasn't the test that she had expected or prepared for. Instead of being assessed on the wards, she'd faced a challenge that was deeper and closer to home. It wasn't about treating her patients like family, but treating her family like a patient. Which, it turned out, was the far more trying task.

"I wish I didn't have to leave home and start residency," she told Akiva that week. "I feel guilty. I feel like I should stay and keep watching him."

"No, this is the deal you made with society," Akiva told her. "You take care of other people's dads, and they take care of yours. It's only right that you give back what you got."

Twenty-One

SAM, May 2020

In that other, non-coronavirus reality, Sam had been meant to spend the spring traveling with his family. The email arrived on his phone on a Thursday: Your boarding pass is ready. That his itinerary was actually a bike ride to Bellevue and a twelve-hour shift seemed like a cruel joke designed by either God or Delta Airlines.

His mom had started planning their family trip when Sam was home the spring before, during his third year of medical school. She wanted the whole family to get away together before Sam started residency. It seemed like the end of something—unencumbered, precareer family time. They decided on the Grand Canyon and booked their tickets in December.

"How do we feel about the nighttime stargazing tour?" his mom had asked at some point.

They felt positively. They were supposed to hike dusty West Coast trails and swim in Lake Bryce. Jeremy would come too, and then he and Sam planned to fly to San Francisco, where Jeremy would officiate a wedding for friends. Now all those mental images felt like a tease, along with that airline email.

Sam knew he'd made the right choice, signing up to work

at Bellevue, but the exhaustion also wore on him. Sometimes Jeremy, an avid member of his own media company's workplace union, grew incensed on Sam's behalf when they talked about how long the hospital shifts could run. The work didn't even end when Sam left in the evenings. He still worried about his patients and checked their records on his phone from home, what was called "chart stalking," to make sure they were stable.

Sam tried to call his family at night, but was sometimes too tired. By the time he got back to the apartment at 8:00 or 8:30, he was nearly ready to collapse in bed. Laurie and Neil tried not to pester him with all their worries. But as they followed headlines and cable TV segments from their home in Ohio, it made them nervous that Sam was at the epicenter of it all. Slowly they were adjusting to their own new routines, months stripped of any trips to synagogue, the movies, or the local nature center. This lockdown reminded them of Sundays during their youth in small southern towns when all the businesses shut down—it was just weeks and weeks of straight Sundays. One evening Jeremy sent them a photo of Sam standing on a typically bustling SoHo corner that was now ghostly quiet. It landed on their phones like a gut punch. It was a visceral reminder of the gap between the New York City they knew and this hazardous streetscape the boys now occupied.

Adding to the woes in their household, Sam's grandma fell and broke her wrist that spring. Laurie jumped in her car and sped over to her living facility, where she had to wrangle special permission just to gain entry to her mom's apartment. Laurie decided not to call Sam that evening because she didn't want to scare him while she waited for updates from the doctor.

Late that night, the independent living facility told his grandmother that she wouldn't be allowed out of her room for the next two weeks, in case she'd been exposed to anything in the hospital. Laurie reminded her mom that at least she was healthy, that was most important. But Grandma wasn't so sure. What was her health worth, she wanted to know, if she couldn't make it to bridge night?

Sam's grandma kept lightly pressing for him to come back to Ohio right after his Bellevue assignment—"It'd be great if you came home!"—which cracked Sam up.

"Let's check the facts on that," he told her. His parents were both in their sixties, and his grandma was ninety. If Sam came home and any one of them so much as coughed, he'd instantly spiral into a vortex of shame.

Back at Bellevue, Sam had his own elderly patients to care for. One of them had a roundabout hospital stay: he was hospitalized with Covid and spent three weeks in the ICU, got discharged home, had a massive heart attack, and was readmitted to the ICU. In the last days of his second stay, he was moved from the ICU to Sam's care.

The patient was delirious from his weeks of critical care. He kept confusing the speech therapist with his daughter. His family assured Sam that he spoke some English, but in his post-ICU delirium he seemed only able to pick up what was said to him in Spanish.

Sometimes the man had flashes of lucidity and asked to talk with his family. Sam had written the phone numbers on a Post-it note, but the patient's fingers were too weak to dial them on his own. He cried when he heard his daughter's voice on the line, and then again when he heard his wife's.

One afternoon he asked to speak to a person whose name Sam didn't recognize. "He's a little delirious," Sam told the daughter over the phone. "Do you know who this person is?"

"Oh!" The daughter laughed. "That's his car mechanic."

Sam wondered if this could be a further sign of the patient's delirium. But before he could ask, the daughter told him that was a logistical request—the mechanic had his car. Maybe his mental fog was clearing up after all.

After a few days, the patient began to seem more oriented to the hospital. Faces started to look familiar; instructions from the nurses made more sense. Soon he would be ready for discharge to a skilled nursing facility. The day before he left, his daughter called Sam with a last request. "Can you make sure he calls his wife tonight?" she asked. "It's her birthday. He's been so confused, and I want to make sure he doesn't forget."

Sam had already thrown out his mask, but he stopped by the man's room and greeted him from the door. "It's been great taking care of you," he said. "I'm glad you're doing so much better."

But Sam suddenly realized he was interrupting a call. The patient was on the phone with his wife, wishing her a happy birthday. Sam figured this was as good a sign as any of his recovery.

"When my strength is back, I'm going to walk back here and thank you all," the man called from his bed.

Sam told him, with a laugh, that whatever he did, he shouldn't come back to the hospital.

On the morning of Sam's last frontline shift, he walked out of his apartment into a better-than-summer Sunday. It was 64 degrees, breezy, with a powder-blue sky. Sam was tired, but at least his work at Bellevue was adjusting back to normal. He was starting to see textbook cases of acute alcoholic liver failure and heart failure from atrial fibrillation. By June there would be only fifty or so Covid patients in the Bellevue wards—down from over three hundred and fifty in April—and a dozen in the ICU.

With the pace slowing, Sam kept thinking back to a piece of advice he'd been given in middle school. A teacher had told him he had to "learn how to learn." That was his goal in the hospital in these final days, thinking about the type of learning he'd done in the Covid wards and what he still had to learn, heading into residency.

As Sam strolled, he passed people with masks secured on their faces and others flouting the safety and social distancing norms.

The city's lockdown was lifting, and its residents were doing the dance of calculating their risk levels—how many friends to see, how many feet to keep apart?

Sam and Jeremy had begun to joke that all of America suddenly understood what it was like to be a gay man in the late 1990s, stories that they had wrestled with for years. People finally got how tiring and stressful it was to perform a constant social choreography—calibrating your every action to personal risk tolerance and possible threat to the people around you. No one wanted to expend the mental and physical energy needed for precautionary measures, especially people who felt young, healthy, and just wanted to live their lives like normal. But in a pandemic, no health choices were personal.

"Part of me has my arms crossed, nodding smugly," Sam said one afternoon. "Everyone understands what it's like to navigate social decisions and decide your risk threshold. It's exhausting." All the caution and vigilance that he had preached for years had become the new normal.

He kept thinking about the testing for sexually transmitted infections that he did in college. He thought about the stories he read of people living in gay communities in the late 1980s and early '90s, when sex could be a death wish and the virus a death sentence. He thought about coming out in college and worrying that having condomless sex meant you'd get HIV and die. The trauma was so ingrained. It was remarkable, to Sam, how simple prevention could be, and at the same time how impossibly difficult. Certain people would have to bear the burden of being boring and responsible, and there would always be a level of stigma associated with being "no fun." It was somehow not that hard to cordon your personal health choices away from public health sermons—even when the stakes were potentially fatal, when the message was *Wear a condom or you die*, or *Wear a mask or you get Covid*.

Sam could be like a biblical prophet, shaking everyone by the shoulders and telling them to wear their masks. He could yell about the codes that he'd witnessed. But what good would it do? It was never just about Covid. People would bike without their helmets on. People would get into Ubers without buckling their seat belts. People would drink too much, and eat too much sugar, and pop too many pills. People would fuck without condoms. They would fuck without PrEP. That was their nature. Our nature.

And for so many people, life made it all but impossible to prioritize health and safety. Like some of the patients Sam saw at Bellevue, who were only partly focused on their fears of coronavirus infection because they had so many other immediate needs: food, work, shelter. But then there would always be the thoughtless people too. Those with all the resources to make the right choices who picked the risks of convenience anyway, and put others under threat.

After his Bellevue weeks wrapped up, Sam took a Covid test, which felt to him like getting water up his nose mid-swim. Once he had received his negative result and quarantined for two weeks, he and Jeremy drove ten hours to Ohio to visit Sam's family.

Back in Cincinnati, the couple stumbled through the rituals of their strange new Covid world. They had their first meal at a restaurant, which was operating at reduced capacity and had placed garish displays of wine bottles on all the empty tables. They did curbside pickup and got a Panera lunch to eat with Sam's grandma. Sam's parents finally had the chance to meet Jeremy's, though it was over a video call—not exactly the celebratory dinner party they'd once envisioned. It was a gathering that compounded the distinct awkwardness of meeting a significant other's family with the universal awkwardness of any encounter on Zoom. But at least it was devoid of all the pressures bred by Match Day, hotel reservations, and flights to New York. Soon after, their mothers decided they would take a Psalms class together, offered virtually through Sam and Jeremy's queer New York congregation.

One night Jeremy visited with his family by attending Zoom services for the Texas synagogue where his mom worked as a cantor. Jeremy's brother, a transgender man, was giving a virtual sermon about allyship and empathy. It was the type of reunion made possible only in pandemic times: celebrating Shabbat together while Jeremy was at Sam's home in Ohio and his own family was scattered across Fort Worth, Nashville, and New York.

Meanwhile, Sam was spending his days unwinding and preparing for the start of residency. His intern year would begin with two weeks treating cancer patients, and then two weeks in a medical mystery diagnosis unit.

Sam was looking forward to celebrating Pride as they settled back home in New York, though he knew the city's festivities would be sapped of their typical drunken bliss. There would be no Friday-night parties at cramped Manhattan apartments and group chats with friends of friends all trying to line up the best weekend plans.

It was the first time he woke up on Pride Day without FOMO or raging body dysmorphia. None of that seemed to matter this year. It wasn't a holiday about brunch and party tickets. He thought about the people who fought to take up space when they weren't meant to, the politics of survival.

Sam pulled on a black tank top, preparing for the scorching June sun, and joined a mass of people marching from Sixth Avenue to the Stonewall Inn on Christopher Street, then back to Washington Square Park. People danced, sang, and sported fanny packs. Mostly their chants were connected to Black Lives Matter, the protesters being cognizant that this wasn't Pride at any normal time.

There were shouts of "I can't breathe!" The crowd called the names of those killed by police violence: Breonna Taylor. George

Floyd. Then, the name of a young Afro-Latinx transgender woman who died at Rikers in June 2019. "Layleen Polanco!" they yelled. At one point, someone called out: "White bodies to the front!" White protesters moved up through the crowd, so they would be first to encounter the police officers.

In past years, Sam and his friends had speculated: What would Pride look like if it wasn't so corporate? Now they had their answer, a day that was genuinely about politics and principles. One that wasn't just designed for white and wealthy New Yorkers. Sam had been involved in queer advocacy for years, but normally it was in professional settings. This was more raw. He moved through the familiar streets of his historically gay neighborhood, which was now lined with police barricades.

As the crowd wound its way toward Stonewall, Sam thought about a patient he had met a few weeks ago. The woman, in her late thirties, had arrived at Bellevue with a pain crisis from sickle cell anemia. The day she was admitted was a peak Covid day. Most of the others in Sam's care had that signature positive swab. Sam noticed that the patient with sickle cell seemed intimately familiar with Bellevue's layout. She didn't ask for anyone's help finding the microwave available for patient use. It seemed like she had spent a fair amount of time in the hospital.

Sometime that day, she had reminded Sam that sickle cell disease affected Black patients disproportionately. The incidence rate was seventy-three cases for every thousand Black babies born, and three cases per thousand for white. And not as many folks knew about it, the woman told Sam, because no one made pink ribbons to raise awareness. It didn't have its own parade.

"We don't get pink ribbons, we don't get 5Ks," she said. "Y'all don't care because it's a Black disease."

Sam had been thinking about her ever since. He was thinking about which diseases are deemed deserving of parades, or even of newspaper coverage.

During the AIDS crisis, it wasn't until wealthy white celebrities began dying that the government started to invest serious resources in its response. It took four years for then-president Reagan to even mention the disease in public. During Covid, there were people who started to let up on their panic once they realized who the virus was killing first. They pressed on with holiday travel, birthday dinners, and house parties without thought for all the more vulnerable caught up in those crosshairs. It was a disease of the old and sick, but also a disease of the Black, Hispanic, and poor.

The racial disparities that the pandemic spotlighted had been evident to Sam and his coworkers long before Covid—it was impossible not to see inequities in health while working at Bellevue. But the pandemic made manifest just how deep those inequities ran, and just how much it took to mobilize public response.

In the weeks since Sam saw that patient with sickle cell, the nation had erupted. People were rallying in protest against the racism that coursed through American veins, through its medical system, and through its months of pandemic. Millions were filling the streets to name those injustices. This was a crowd, a city, a country, marching into the addled territory of the unknown. After all the suffering they'd witnessed, with all these new scars, they were reminded of what they hadn't lost: feet on pavement, arms raised skyward, muffled voices overlapping in the hot city air.

Acknowledgments

Since I arrived at the *New York Times* two years ago, I've been surrounded by examples of what it means to be a top-notch reporter and colleague.

I owe so much to Katie Kingsbury, who hired me, assigned my first story, and has been the greatest source of encouragement and inspiration ever since; you are a role model for me and for so many. Thank you to Alex Kingsbury and Lauren Kelley, who taught me to edit with a sharp eye and sense of humor; to members of opinion and the editorial board—Michelle Cottle, Brent Staples, Binyamin Appelbaum, Greg Bensinger, Farah Stockman, Jesse Wegman, Nick Fox—who dazzled me with their brilliance and welcomed me with such warmth. Thank you to Nick Kristof, who was my hero growing up and then became my advocate and best source of story ideas.

I'm so grateful for my coworkers who quickly became the dearest of friends: Jeneen Interlandi, thank you for the smoothies and walks and words of life wisdom that kept me sane; Mara Gay, you are a powerhouse writer and were the first to show me that the fiercest journalists are also the most bighearted. You taught me so much about empathy and commitment—to many more runs into the ocean.

Thank you to my editors, who took a chance on me from the

moment I arrived: to the incredible Rachel Dry, the only person who could draw a crowd for Zoom standup comedy; to Francesca Donner, who rules Team Gender with unending dynamism; to Alan Burdick, Michael Roston, and Kate Phillips, who shaped my stories and made them sing.

This book was born out of an exchange with Sam Freedman, who has been teaching me how to report since high school. Thank you for believing in this story before anyone; your depth as a journalist is matched only by your devotion as a mentor. Thank you to Anne Fadiman, whose soulful lessons on writing I revisit daily—I hope this book had sufficient left clavicles. I owe so much to the singular Rebecca Wallace-Segall, who turned an awkward seventh-grade reader into a journalist at Writopia.

Thank you to Sam, Gabriela, Iris, Jay, Elana, and Ben for opening up your lives and hearts to me, and teaching me so much about medicine, equity, and courage. I emerged from each one of our many conversations a better person, and I am excited to keep watching what you do for your patients, your city, and your communities.

I'm so grateful to everyone at HarperCollins for guiding me through this process. Noah Eaker, you are the dream book editor—as genius, as you are patient with a newbie, as kind as you are insightful. Thank you for seeing this book from the beginning, wrestling with it every step of the way, and making it all a joy. Thank you to Kate D'Esmond, Becca Putman, Mary Gaule, Kate Childs, and Milan Bozic for all your remarkable work on this project, and for making me feel like I was in such good hands. Thank you to Miranda Ottewell for your excellent copyedits.

Thank you to my agents, Dave Larabell and Cindy Uh, for your thoughtfulness and your savvy, and for championing this book from our earliest calls. The time and enthusiasm you devoted to it blew me away. Thank you to the angelic Carrie Frye, who is delightful and incisive in equal measure; I left every one of our discussions feeling better about the writing and the world.

My first reporting on medical schools came out of a phone call from Adam Beckman, whose notebook of observations is going to change the field. Thank you to Eric Cervini, David Shimer, and Sopan Deb, who passed on every bit of wisdom from their own brilliant books. Thank you to the trailblazers whose writing and conversations sharpened my thinking and challenged my perspective: Dr. Atul Gawande, Dr. Rana Awdish, Dr. Jessi Gold, and David Oshinsky.

Thank you to all the doctors who took time out of their long shifts to read drafts, pressure-test ideas, and push for deeper thinking: Raquel Sofia Sandoval, Margaret Zhang, Adaira Landry, Gifty Kwakye, Onyekachi Otugo, David Edelman, Hashem Zikry, Maxine Dexter, Lash Nolen, Leah Schwartz, Lauren Tronick, Adrian Chiem, Sami Stein, Danielle Cameron, Colleen Farrell, Luis Seija, Melissa Hill. Thank you to the inimitable Sophie Cain Miller, whose rigorous medical fact-checking was a godsend, and to Kate Wheeling, who pored over every bit of historical information.

I couldn't have written this book without my friends, who provided me with encouragement and ice cream throughout the process. Thank you to Emily Kassie, the only person who would dive into a draft and call hours later with "show notes," then spend a whole day at Bellevue to photograph the cover. You're an exceptional journalist, and an even better friend. To Corey Malone-Smolla, who lets me be one more sibling in her big, loving family; to Catherine Wang and Jenny Allen, my highest-flow people, who tell the best stories, pick the best drafts, and show up with such endless support; to Gina Starfield, Sophie Paci, Lara Sokoloff, and the Elmhurst crew, whose company and friendsgivemukkahs make everywhere home; to Chris Moates, who is as giving in his edits as in every other area of life; to my first and favorite co-editor, Geng Ngarmboonanant. To Ally Daniels, Kristi Murray, Amelia Nierenberg, Natalie Epstein, Katie Galbraith, and Gabe Fisher, for being guardian angels and book spirit guides. To all my friends who brought so much light into this distanced year.

Thank you to Tim Krupa, my partner-in-pod, for filling each day with laughter, love, frozen mangoes, and all the brightest ideas—including for the title of this book. You made it all worthwhile. To my brother, Coby Goldberg, objectively the best person: thanks for cracking me up, editing my drafts, never tiring of a "bit," never turning down a lunch special, and gently reminding me that if I don't read *Anna Karenina* before age thirty, I probably never will. To my dad, J.J. Goldberg: spending late nights writing with you made me want to become a reporter, and you've been my source of guidance, love, and lousy jokes ever since. To Mike Glassberg, for bringing so much happiness to the family. And to my mom, Shifra Bronznick, my rock and truth-teller, who has always modeled the kind of generosity, boldness, and care I didn't even know was possible. There's an impossibly large web of people you've touched, and I'm lucky to be in the thick of it. I know that *your* mom liked to say, "Everybody loves Shifra," and while this is true, I know that no one could love you more than I do.

A Note on Sources

Shortly after reporting for the *New York Times* on the medical schools that accelerated their graduations because of Covid-19, I began speaking with the group of six early graduates who appear in this book. I spent hundreds of hours interviewing the doctors as well as their family members, partners, classmates, and coworkers. I visited their hospitals and, in some cases, reviewed their photos, videos, and notes. I also spoke to dozens of other doctors who worked with Covid-19 patients, both in and outside New York City, and to researchers focused on health equity. I benefited from the generosity of academics and journalists focused on the history of medicine who shared insights from their decades of work.

I relied on reporting about the coronavirus pandemic that appeared in many publications but especially the *New York Times*, *STAT News*, the *New Yorker*, Reuters, NPR, and *ProPublica*. I used Covid-19 data from the Centers for Disease Control and Prevention. I also relied on the extensive historical research on American medical schools conducted by scholars such as Paul Starr, Kenneth Ludmerer, and David Oshinsky.

Here are the sources that were most useful for each chapter.

Introduction

In writing the introduction, I relied primarily on interviews with Sam, along with his classmate Jessica, who was also present on the early graduates' first day at Bellevue. I also made use of diversity statistics from the Association of American Medical Colleges, data on median weekly earnings from the US Bureau of Labor Statistics, reporting in *STAT News*, and data on Covid-19 from the Centers for Disease Control and Prevention and the Economic Policy Institute. The following published sources were especially useful:

Alsan, Marcella, Owen Garrick, and Grant C. Graziani. "Does Diversity Matter for Health? Experimental Evidence from Oakland." *NBER Working Paper No. 24787* (June 2018). https://www.nber.org/papers/w24787.

Starr, Paul. *The Social Transformation of American Medicine*. New York: Basic Books, 2017.

Chapter One

In writing this chapter, I relied primarily on interviews with Sam; his partner, Jeremy; and Sam's parents, Laurie and Neil. I also interviewed Dr. Steven Abramson, the vice dean for education at NYU Medical School, and Alison Whelan, the chief medical education officer of the Association of American Medical Colleges. I made use of data from the Association of American Medical Colleges and reporting on Covid-19 in the *New York Times*, CNN, Reuters, the *New Yorker*, the *Guardian*, and the *Atlantic*. The following published sources were especially useful:

Herman, Benjamin, Rhonda J. Rosychuk, Tracey Bailey, Robert Lake, Olive Yonge, and Thomas J. Marrie. "Medical Students and Pandemic Influenza." *Emerg Infect Dis* 13, no. 11 (2007): 1781–83. https://dx.doi.org/10.3201/eid1311.070279.

Schwartz, Christine C., Aparna S. Ajjarapu, Chris D. Stamy, and Debra A. Schwinn. "Comprehensive History of 3-Year and Accelerated US Medical School Programs: A Century in Review." *Med Educ Online* 23, no. 1 (2018): 1530557. https://doi.org/10.1080/10872981.2018.1530557.

Chapter Two

In writing this chapter, I relied primarily on interviews with Gabriela, her mother, and her partner, Jorge. I also interviewed Paul Starr, who has studied the history of American medical schools. I made use of data from the Association of American Medical Colleges and the Kaiser Family Foundation, the Carnegie Foundation's digital archive, and the annual reports of the president of Harvard University in 1869 and 1870. The following published sources were especially useful:

Campbell, Kendall M., Irma Corral, Jhojana L. Infante Linares, et al. "Projected Estimates of African American Medical Graduates of Closed Historically Black Medical Schools." *Journal of the American Medical Association* 3, no. 8 (2020): e2015220. https://doi.org/10.1001/jamanetworkopen.2020.15220.

Cooper, Richard. "Medical Schools and Their Applicants: An Analysis." *HealthAffairs* 22, no. 4 (2003). https://doi.org/10.1377/hlthaff.22.4.71.

Ludmerer, Kenneth. *Let Me Heal*. New York: Oxford University Press, 2015.

Simmenroth-Nayda, Anne, and Yvonne Görlich, "Medical School Admission Test: Advantages for Students Whose Parents Are Medical Doctors?" *BMC Medical Education* 15, no. 81 (2015). https://doi.org/10.1186/s12909-015-0354-x.

Starr, Paul. *The Social Transformation of American Medicine*. New York: Basic Books, 2017.

Stone, James. "The Relations of the Massachusetts Medical Society to the Public." *Boston Medical Surgery Journal* 190 (1924): 1005–12. https://doi.org/10.1056/NEJM192406121902401.

Chapter Three

In writing this chapter, I relied primarily on interviews with Iris and her partner, Benjamin. I also interviewed Dr. Rana Awdish and Dr. Uché Blackstock. I made use of reporting in the *New York Times*, particularly by Aaron Carroll and Austin Frakt, and the digital archives of the US Holocaust Museum. The following published sources were especially useful:

Awdish, Rana. *In Shock*. New York: St. Martin's Press, 2017.

Emanuel, Ezekiel, and Linda Emanuel. "Four Models of the Physician-Patient Relationship." *Journal of the American Medical Association* 267, no. 16 (1992): 2221–26. https://doi.org/10.1001/jama.1992.03480160079038.

Gawande, Atul. "Whose Body Is It Anyway?" *New Yorker*, October 4, 1999. https://www.newyorker.com/magazine/1999/10/04/whose-body-is-it-anyway.

Kaba, R., and P. Sooriakumaran. "The Evolution of the Doctor-Patient Relationship." *International Journal of Surgery* 5, no. 1 (2007): 57–65. https://doi.org/10.1016/j.ijsu.2006.01.005.

Katz, Jay. *The Silent World of Doctor and Patient*. New York: Free Press, 1984.

Lown, Bernard. *The Lost Art of Healing: Practicing Compassion in Medicine*. New York: Random House, 1996.

Metcalfe, D. "Whose Data Are They Anyway?" *British Medical Journal* (Clinical Research Ed.) 292 (1986): 577–78. https://doi.org/10.1177/0038038504039366.

Moreno, Jonathan D., Ulf Schmidt, and Steve Joffe. "The Nuremberg Code 70 Years Later." *Journal of the American Medical Association* 318, no. 9 (2017). https://doi.org/10.1001/jama.2017.10265.

Roeland, Eric, Julia Cain, Chris Onderdonk, Kim Kerr, William Mitchell, and Kathryn Thornberry. "When Open-Ended Questions Don't Work." *Journal of Palliative Medicine* 17, no. 4 (2014): 415–20. https://doi.org/10.1089/jpm.2013.0408.

Rosenthal, Elisabeth. *An American Sickness: How Healthcare Became Big Business and How You Can Take It Back*. New York: Penguin, 2017.

Schneider, Carl. *The Practice of Autonomy*. New York: Oxford University Press, 1998.

Tingley, Kim. "Trying to Put a Value on the Doctor-Patient Relationship." *New York Times Magazine*, May 16, 2018. https://www.nytimes.com/interactive/2018/05/16/magazine/health-issue-reinvention-of-primary-care-delivery.html.

Vollmann, J., and R. Winau. "Informed Consent in Human Experimentation before the Nuremberg Code." *British Medical Journal* 313, no. 7070 (1996): 1445–49. https://doi.org/10.1136/bmj.313.7070.1445.

Weissman, Joel, and Joseph Betancourt. "Resident Physicians' Preparedness to Provide Cross-Cultural Care." *Journal of the American Medical Association* 294, no. 9 (2005): 1058–67. https://doi.org/10.1001/jama.294.9.1058.

Wells, Lindsay, and Arjun Gowda. "A Legacy of Mistrust." *Proceedings of UCLA Health* 24 (2020).

Chapter Four

In writing this chapter, I relied primarily on interviews with Gabriela; her partner, Jorge; and her mother; as well as their photos. I made use of reporting in the *New York Times*, by Sheri Fink, J. David Goodman, William K. Rashbaum, Jeffrey C. Mays, and Joseph Goldstein; in *ProPublica*, by Charles Ornstein, Joe Sexton, and Joaquin Sapien; in *Vox*, by Cameron Peters; and in the *Atlantic*, by Ed Yong. I also drew on Covid-19 data from the Centers for Disease Control and Prevention.

Chapter Five

In writing this chapter, I relied primarily on interviews with Jay and her mother, along with Jay's emails and photographs.

Rosenthal, Elisabeth. *An American Sickness: How Healthcare Became Big Business and How You Can Take It Back*. New York: Penguin, 2017.

Chapter Six

In writing this chapter, I relied primarily on interviews with Elana, Ben, and Ben's sister Jenny, along with their photos.

Chapter Seven

In writing this chapter, I relied primarily on interviews with Sam, as well as visits to Bellevue. I interviewed Dr. Douglas Bails, Bellevue's chief of medicine, AIDS expert Dr. Fred Valentine, historian Sandra Opdyke, and David Oshinsky, author of *Bellevue*. I made use of the *New York Times*' reporting during the AIDS crisis, particularly the work of Bruce Lambert, Ronald

Sullivan, and Lawrence K. Altman, and the digital archives of the American Medical Association. The following published sources were especially useful:

Defoe, Daniel. *A Journal of the Plague Year.* New York: Penguin Classics, 2003.

Ofri, Danielle. "Imagine a World Without AIDS." *New York Times*, July 27, 2012. https://www.nytimes.com/2012/07/28/opinion/imagine-a-world-without-aids.html.

———. "Pas de Deux," in Lee Gutkind, ed., *Becoming a Doctor: From Students to Specialists, Doctor-Writers Share Their Experiences.* New York: W.W. Norton, 2010.

Oshinsky, David. *Bellevue.* New York: Penguin Random House, 2016.

Rockmore Angoff, Nancy. "Do Physicians Have an Ethical Obligation to Care for Patients with AIDS?" *Yale Journal of Biology and Medicine* 64, no. 3 (1991): 207–46. https://www.ncbi.nlm.nih.gov/pmc/articles/PMC2589324/.

Yin, Sophia. "Physicians' Duty to Treat in a Pandemic: A Code of Ethics Approach." *Harvard Medical Student Review*, June 21, 2020. https://www.hmsreview.org/covid/physicians-duty-to-treat.

Chapter Eight

In writing this chapter, I relied primarily on interviews with Iris, Benjamin, and Elana. I interviewed historian Kenneth Ludmerer and medical student Melissa Hill. I also made use of reporting in the *New York Times*, particularly by Mariel Padilla and Sanya Dosani, and in *STAT News*, by Lauren Joseph. The following published sources were especially useful:

Heschel, Abraham Joshua. *The Sabbath.* New York: Farrar, Straus and Giroux, 2005.

Hughes, Emily. "July Effect? Maybe Not." *Canadian Medical Association Journal* 189, no. 32 (2017). https://doi.org/10.1503/cmaj.1095466.

Ludmerer, Kenneth. *Let Me Heal.* New York: Oxford University Press, 2015.

Chapter Nine

In writing this chapter, I relied primarily on interviews with Jay and her coworkers, along with Jay's notes and photos and a visit to her hospital. I interviewed other Covid-19 doctors, including Dr. Manish Garg, Dr. Joshua Davis, Dr. Megan Ranney, Dr. Hashem Zikry, Dr. Luis Seija, Dr. Lakshman Swamy, Dr. Mizuho Morrison, and Dr. Richina Bicette; AIDS expert Dr. Fred Valentine; health equity researchers Dr. Marcella Alsan, Dr. Louis Penner, Dr. Uché Blackstock, Dr. Susan Persky, Dr. Nao Hagiwara, Cecile Yancu; and freelance journalist Patrice Peck. I made use of reporting in NBC, Reuters, the *New York Times*, NPR, *STAT News*, ABC News, and PBS as well as Covid-19 data from the Centers for Disease Control and Prevention. The following published sources were especially useful:

Alpert, Alison B., Eileen E. Cichoskikelly, and Aaron D. Fox. "What Lesbian, Gay, Bisexual, Transgender, Queer and Intersex Patients Say Doctors Should Know and Do: A Qualitative Study." *Journal of Homosexuality* 64, no. 10 (2017): 1368–89. https://doi.org/10.1080/00918369.2017.1321376.

Alsan, Marcella, Owen Garrick, and Grant C. Graziani. "Does Diversity Matter for Health? Experimental Evidence from Oakland." *National Bureau of Economic Research*. Working Paper 24787, June 2018. https://doi.org/10.3386/w24787.

Alsan, Marcella, and Marianne Wanamaker. "Tuskegee and the Health of Black Men." *Quarterly Journal of Economics* 133, no. 1 (2018). https://doi.org/10.3386/w22323.

Bower, Julie K., Pamela J. Schreiner, Barbara Sternfeld, Core E. Lewis. "Black-White Differences in Hysterectomy Prevalence." *American Journal of Public Health* 99, no. 2 (2009): 300. https://doi.org/10.2105/AJPH.2008.133702.

Calabrese, Sarah K., Valerie A. Earnshaw, Kristen Underhill, Douglas S. Krakower, Manya Magnus, Nathan B. Hansen, Kenneth H. Mayer, Joseph R. Betancourt, Trace S. Kershaw, and John F. Dovidio. "Prevention Paradox: Medical Students Are Less Inclined to Prescribe HIV Pre-Exposure Prophylaxis for Patients in Highest Need." *Journal of the International AIDS Society* 21, no. 6 (2018): e25147. https://doi.org/10.1002/jia2.25147.

Chapman, Elizabeth N., Anna Kaatz, and Molly Carnes. "Physicians and Implicit Bias: How Doctors May Unwittingly Perpetuate Health Care Disparities." *Journal of General Internal Medicine* 28, no. 11 (2013): 1504–10. https://doi.org/10.1007/s11606-013-2441-1.

Corbie-Smith, G. "The Continuing Legacy of the Tuskegee Syphilis Study: Considerations for Clinical Investigation." *American Journal of the Medical Sciences* 317, no. 1 (1999): 5–8. https://doi.org/10.1097/00000441-199901000-00002.

Esnaola, Nestor F., and Marvella E. Ford. "Racial Differences and Disparities in Cancer Care and Outcomes: Where's the Rub?" *Surgical Oncology Clinics of North America* 21, no. 3 (2012): 417–37. https://doi.org/10.1016/j.soc.2012.03.012.

Feinglass, Joe, Cheryl Rucker-Whitaker, Lee Lindquist, Walter J. McCarthy, and William H. Pearce. "Racial Differences in Primary and Repeat Lower Extremity Amputation: Results from a Multihospital Study." *Journal of Vascular Surgery* 41, no. 5 (2005): 823–29. https://doi.org/10.1016/j.jvs.2005.01.040.

Frazer, Somjen M. "LGBT Health and Human Services Needs in New York State." *Empire State Pride Agenda Foundation*, 2009.

Greenwood, Brad N., Seth Carnahan, and Laura Huang. "Patient-Physician Gender Concordance and Increased Mortality among Female Heart Attack Patients." *Proceedings of the National Academy of Sciences of the United States of America* 115, no 34 (2018): 8569–74. https://doi.org/10.1073/pnas.1800097115.

Greenwood, Brad N., Rachel R. Hardeman, Laura Huang, and Aaron Sojourner. "Physician-Patient Racial Concordance and Disparities in Birthing Mortality for Newborns." *Proceedings of the National Academy of Sciences of the United States of America* 117, no. 35 (2020): 21194–200. https://doi.org/10.1073/pnas.1913405117.

Hoberman, John. *Black and Blue: The Origins and Consequences of Medical Racism*. Los Angeles: University of California Press, 2012.

Hoffman, Kelly M., Sophie Trawalter, Jordan R. Axt, and M. Norman Oliver. "Racial Bias in Pain Assessment and Treatment Recommendations, and False Beliefs about Biological Differences between Blacks and Whites." *Proceedings of the National Academy of Sciences of the United States of America* 113, no. 16 (2016): 4296–301. https://doi.org/10.1073/pnas.1516047113.

Malhotra, Jyoti, David Rotter, Jennifer Tsui, Adana A.M. Llanos, Bijal A. Balasubramanian, and Kitaw Demissie. "Impact of Patient-Provider

Race, Ethnicity, and Gender Concordance on Cancer Screening: Findings from Medical Expenditure Panel Survey." *Cancer Epidemiology, Biomarkers and Prevention* 26, no. 12 (2017): 1804–11. https://doi.org/10.1158/1055-9965.EPI-17-0660.

Newkirk, Vann R. "A Generation of Bad Blood." *The Atlantic*, June 17, 2016. https://www.theatlantic.com/politics/archive/2016/06/tuskegee-study-medical-distrust-research/487439/.

Persky, Susan, Kimberly A. Kaphingst, Vincent C. Allen Jr., and Ibrahim Senay. "Effects of Patient-Provider Race Concordance and Smoking Status on Lung Cancer Risk Perception Accuracy among African-Americans." *Annals of Behavioral Medicine* 45, no. 3 (2013): 308–17. https://doi.org/10.1007/s12160-013-9475-9.

Sabin, Janice A., Rachel G. Riskind, and Brian A. Nosek. "Health Care Providers' Implicit and Explicit Attitudes Toward Lesbian Women and Gay Men." *American Journal of Public Health* 105, no. 9 (2015): 1831–41. https://doi.org/10.2105/AJPH.2015.302631.

Skloot, Rebecca. *The Immortal Life of Henrietta Lacks*. New York: Crown Publishers, 2010.

Traylor, Ana H., Julie A. Schmittdiel, Connie S. Uratsu, Carol M. Mangione, Usha Subramanian. "Adherence to Cardiovascular Disease Medications: Does Patient-Provider Race/Ethnicity and Language Concordance Matter?" *Journal of General Internal Medicine* 25, no. 11 (2010): 1172–77. https://doi.org/10.1007/s11606-010-1424-8.

Washington, Harriet A. *Medical Apartheid*. New York: Anchor Books, 2007.

Chapter Ten

In writing this chapter, I relied primarily on interviews with Gabriela and Ben, as well as on visits to their hospitals. I made use of reporting in Reuters and the *New York Times*, as well as data from the Centers for Disease Control and Prevention. The following published sources were especially useful:

Bartik, Alexander W., Marianne Bertrand, Zoë B. Cullen, Edward L. Glaeser, Michael Luca, and Christopher T. Stanton. "How Are Small Businesses Adjusting to Covid-19?" *National Bureau of Economic Research*, Working Paper 26989, April 2020. https://doi.org/10.3386/w26989.

Curtis, J. Randall, Donald L. Patrick, Ellen S. Caldwell, et al. "Why Don't Patients and Physicians Talk About End-of-Life Care?" *Journal of the American Medical Association Internal Medicine* 160, no. 11 (2000): 1690–96. https://pubmed.ncbi.nlm.nih.gov/10847263/.

Davis, Matthew A., Brahmajee K. Nallamothu, Mousumi Banerjee, and Julie P. W. Bynum. "Patterns of Healthcare Spending in the Last Year of Life." *Health Affairs* 35, no. 7 (2016): 1316–23. https://doi.org/10.1377/hlthaff.2015.1419.

Gawande, Atul. *Being Mortal: Medicine and What Matters in the End*. New York: Metropolitan Books, 2014.

Johnson, Kimberly S. "Racial and Ethnic Disparities in Palliative Care." *Journal of Palliative Medicine* 16, no. 11 (2013): 1329–34. https://doi.org/10.1089/jpm.2013.9468.

Nuland, Sherwin. *How We Die: Reflections on Life's Final Chapter*. New York: Vintage Books, 1994.

Pinching, Anthony J., Roger Higgs, and Kenneth M. Boyd. "The Impact of AIDS on Medical Ethics." *Journal of Medical Ethics* 26, no. 1 (2000). https://jme.bmj.com/content/26/1/3.

Sabatino, Charles P. "The Evolution of Health Care Advance Planning Law and Policy." *Milbank Quarterly* 88, no. 2 (2010): 211–39. https://doi.org/10.1111/j.1468-0009.2010.00596.x.

Sadownik, Sara, Hannah James, and David Auerbach. "Serious Illness Care in Massachusetts: Differences in Care Received at the End of Life by Race and Ethnicity." *Massachusetts Health Policy Commission*, September 30, 2020. https://www.mass.gov/doc/policy-brief-serious-illness-care-in-massachusetts-differences-in-care-received-at-the-end-of/download.

Smith, Cardinale B., Sofya Pintova, Kerin B. Adelson, and Jason Parker. "Disparities in Length of Goals of Care Conversations between Oncologists and Patients with Advanced Cancer." *Journal of Clinical Oncology* 36, no. 34 (2018). https://ascopubs.org/doi/10.1200/JCO.2018.36.34_suppl.19.

Smith, Susan L. "War! What Is It Good For? Mustard Gas Medicine," *Canadian Medical Association Journal* 189, no. 8 (2017). https://doi.org/10.1503/cmaj.161032.

Chapter Eleven

In writing this chapter, I relied primarily on interviews with Iris and Elana. I interviewed other Covid-19 doctors, including Dr.

Lynn Fiellin, Dr. Lina Miyakawa, Dr. Doug Bails, Dr. Hilary Fairbrother, Dr. Lakshman Swamy, Dr. William Schaffner, Dr. Joshua Davis, Dr. Megan Ranney, Dr. Hashem Zikry, and Dr. Richina Bicette. I made use of data from the Centers for Disease Control and Prevention.

Chapter Twelve

In writing this chapter, I relied primarily on interviews with Jay, Alicia, and their coworkers, as well as their photos and videos. I made use of data from the Centers for Disease Control and Prevention.

Chapter Thirteen

In writing this chapter, I relied primarily on interviews with Gabriela. I made use of data from the Centers for Disease Control and Prevention.

Chapter Fourteen

In writing this chapter, I relied primarily on interviews with Elana and data from the American Heart Association.

Chapter Fifteen

In writing this chapter, I relied primarily on interviews with Sam. I made use of reporting in the *New York Times*, by Christina Goldbaum, Joseph Goldstein, William K. Rashbaum, and Alan Feuer, and in *The Intercept*, by Alleen Brown.

Chapter Sixteen

In writing this chapter, I relied primarily on interviews with Ben, as well as a visit to his hospital. I interviewed other Covid-19 and palliative-care doctors, including Dr. Atul Gawande, Dr. Hashem Zikry, Dr. Megan Ranney, Dr. Andrew Thurston, Dr. Mizuho Morrison, Dr. Joshua Davis, Dr. Catherine Sarkisian, and Dr. Mitch Wong. I made use of reporting in the *New York*

Times, by Nicholas Kristof, and data from the Centers for Disease Control and Prevention.

Chapter Seventeen

In writing this chapter, I relied primarily on interviews with Jay, her mother, and Jay's coworkers, as well as a visit to her hospital. I interviewed other Covid-19 doctors, including Dr. Amrapali Maitra and Dr. Rana Awdish. The following published sources were especially useful:

Dugdale, L.S. *The Lost Art of Dying: Reviving Forgotten Wisdom*. New York: HarperOne, 2020.

Onishi, Norimitsu. "A Generation in Japan Faces a Lonely Death." *New York Times,* November 30, 2017. https://www.nytimes.com/2017/11/30/world/asia/japan-lonely-deaths-the-end.html.

Wakam, Glenn K., John R. Montgomery, Ben E. Biesterveld, and Craig S. Brown. "Not Dying Alone—Modern Compassionate Care in the Covid-19 Pandemic." *New England Journal of Medicine* 382 (2020). https://doi.org/10.1056/NEJMp2007781.

Chapter Eighteen

In writing this chapter, I relied primarily on interviews with Gabriela and her mother, as well as Gabriela's photos. I used Covid-19 data from the Centers for Disease Control and Prevention and reporting in CNBC and the *New York Times*.

Chapter Nineteen

In writing this chapter, I relied primarily on interviews with Jay and Ben, as well as visits to their hospitals and Jay's photos. The following published sources were especially useful:

Potts, Amanda, and Elena Semino. "Cancer as a Metaphor." *Metaphor and Symbol* 34, no. 2 (2019): 81–95. https://doi.org/10.1080/10926488.2019.1611723.

Sontag, Susan. *Illness as Metaphor and AIDS and Its Metaphors*. New York: Macmillan, 2001.

Chapter Twenty

In writing this chapter, I relied primarily on interviews with Iris, Benjamin, and Elana. The following published source was especially useful:

Park Hong, Cathy. *Minor Feelings*. New York: One World, 2020.

Chapter Twenty-One

In writing this chapter, I relied primarily on interviews with Sam, his parents, Jeremy, and Bellevue's Dr. Doug Bails. I made use of reporting by Mihir Zaveri in the *New York Times* and by Danielle Ofri in the *New Yorker*. I used sickle-cell data from the Centers for Disease Control and Prevention.

About the Author

Emma Goldberg is a reporter at the *New York Times*, writing for sections such as Health and Science, Styles, Gender, National, and Culture, among others. Her cover stories have featured campus techlash, surgeon moms, young women running for office, Hong Kong protests, and low-income medical students. Since the start of the coronavirus outbreak, she has turned her focus to the lives of students, physicians, and nurses battling the pandemic. She is the winner of the Newswomen's Club of New York Best New Journalist Award and the Sidney Hillman Foundation's Sidney Award. Goldberg received her BA at Yale and MPhil at Cambridge University.